Automorphic Forms on $SL_2(\mathbb{R})$ provides an introduction to some aspects of the analytic theory of automorphic forms on $G = SL_2(\mathbb{R})$ or the upper half-plane X, with respect to a discrete subgroup Γ of G of finite covolume. The point of view is inspired by, but does not assume knowledge of, the theory of infinite dimensional unitary representations of G – until the last sections, whose purpose is to introduce this theory and relate it to automorphic forms.

The topics treated include the construction of fundamental domains, the notion of automorphic form on $\Gamma \backslash G$, its relationship with the classical automorphic forms on X, Poincaré series, constant terms, cusp forms, finite dimensionality of the space of automorphic forms of a given type, compactness of certain convolution operators, Eisenstein series, their analytic continuation, unitary representations of G, and the spectral decomposition of $L_2(\Gamma \backslash G)$.

The main prerequisites are some results in functional analysis (reviewed, with references) and some familiarity with the elementary theory of Lie groups and Lie algebras, used only for G and its analytic subgroups.

T0321585

CAMBRIDGE TRACTS IN MATHEMATICS

General Editors
B. BOLLOBAS, F. KIRWAN, P. SARNAK, C. T. C. WALL

130   Automorphic forms on $SL_2(\mathbb{R})$

ARMAND BOREL
Institute for Advanced Study

Automorphic forms on $SL_2(\mathbb{R})$

CAMBRIDGE
UNIVERSITY PRESS

CAMBRIDGE UNIVERSITY PRESS
Cambridge, New York, Melbourne, Madrid, Cape Town, Singapore, São Paulo

Cambridge University Press
The Edinburgh Building, Cambridge CB2 8RU, UK

Published in the United States of America by Cambridge University Press, New York

www.cambridge.org
Information on this title: www.cambridge.org/9780521580496

First published 1997
This digitally printed version 2008

A catalogue record for this publication is available from the British Library

Library of Congress Cataloguing in Publication data
Borel, Armand.
Automorphic forms on $SL_2(R)$ / Armand Borel.
p.   cm. – (Cambridge tracts in mathematics ; 130)
Includes bibliographical references and indexes.
ISBN 0-521-58049-8 (hardbound)
1. Forms, Automorphic.   I. Title.   II. Series.
QA331.B687   1997
515'.9 – dc21   97-6027
                                        CIP

ISBN 978-0-521-58049-6 hardback
ISBN 978-0-521-07212-0 paperback

TO GABY

Contents

Contents

Preface

A more accurate title would be: *Introduction to some aspects of the analytic theory of automorphic forms on* $SL_2(\mathbb{R})$ *and the upper half-plane* X. Originally, automorphic forms were holomorphic or meromorphic functions on X satisfying certain conditions with respect to a discrete group Γ of automorphisms of X, usually with fundamental domain of finite (hyperbolic) area. Later on, H. Maass – and then A. Selberg and W. Roelcke – dropped the assumption of holomorphicity, requiring instead that the functions under consideration be eigenfunctions of the Laplace–Beltrami operator. In the 1950s it was realized (in more general cases) – initially by I. M. Gelfand and S. V. Fomin, and then by Harish-Chandra – that the automorphic forms (holomorphic or not) could be equivalently viewed as functions on $\Gamma \backslash SL_2(\mathbb{R})$ satisfying certain conditions familiar in the theory of infinite dimensional representations of semisimple Lie groups. This led to a new outlook, where the Laplace–Beltrami operator is replaced by the Casimir operator and the theory of automorphic forms becomes closely related to harmonic analysis on $\Gamma \backslash SL_2(\mathbb{R})$. This is the point of view adopted in this presentation. However, in order to limit the prerequisites, no knowledge of representation theory is assumed until the last sections, a main purpose of which is precisely to make this connection explicit. A fundamental role is played throughout by a theorem stating that a function on $SL_2(\mathbb{R})$ satisfying certain assumptions ($\mathcal{Z}$-finite and K-finite) is fixed under convolution by some smooth function with arbitrarily small compact support around the identity element (2.14). This is indeed best understood in the context of representation theory (and valid for a general semisimple group), but the proof in the case of $SL_2(\mathbb{R})$ can and will be described without reference to the general theory.

This topic has now been investigated for well over one hundred years from a variety of angles, and is so rich that there is comparatively not that much overlap between various treatments. Here, we rather single-mindedly explore one direction. As a compensation of sorts, we have tried in Section 0 to give a short (admittedly incomplete) historical introduction that starts from some other more classical topics, indicating how the present point of view arose and how the two are related. However, Section 0 is not referred to later (and may therefore be skipped): whatever from it is used later will be introduced anew. The Introduction ends with a short description of the contents of the various sections; Section 1 lists the prerequisites.

This book is an outgrowth of the notes for a short course given at the Mathematical Institute of the Academia Sinica in spring 1993, upon an invitation of Dr. Ki Keng Lu of the Academia Sinica. A first draft was written by Dr. Xuning Feng, completed by me, translated into Chinese by Dr. Feng and published in the Chinese periodical *Advances in Mathematics* [8]. I have kept much of it, expanding it to more than twice its original size. These notes were the catalytic agent for the present book, but only its immediate predecessor; they came after several series of lectures and uncompleted attempts to write an exposition. A first introduction to the general case was given at the University of Paris in spring 1964, informally published in the last part of [5]. Later I gave courses on automorphic forms on $SL_2(\mathbb{R})$ at MIT in the fall of 1969, at the Universidad Naçional Autonoma in Mexico City in summer 1979, and on automorphic forms on general reductive groups at Yale (fall 1978) and at the Tata Institute of Fundamental Research, Bombay (January–March 1983; unpublished notes by T. N. Venkataramana). Moreover, around 1984, D. Husemoller and I embarked on a project aiming at an exposition of the one-variable case. A certain number of sections were drafted, but this was eventually dropped. In writing up this final version, I have benefited from these various attempts and thank the auditors at these lectures, and D. Husemoller, for their patience and help.

I would also like to thank H. Jacquet for some useful discussions on the later part of the book and a simplification, as well as P. Sarnak, who encouraged me to complete the notes of my Beijing course and offered to have the outcome published in this series. I am grateful to Stephen D. Miller for a very careful reading of the text, thanks to which many typos, minor blemishes, and a blunder have been caught. Thanks are also due to Elly Gustafsson, who speedily and skillfully converted into beautiful AMS-TEX an endless stream of rather unattractive mixtures of handwritten and typewritten drafts.

0 Introduction

0.1 Our framework will be $\mathrm{SL}_2(\mathbb{R})$ and its action on the upper half-plane. Let

$$G = \mathrm{SL}_2(\mathbb{R}) = \left\{ \begin{pmatrix} a & b \\ c & d \end{pmatrix} \,\middle|\, ad - bc = 1,\, a, b, c, d \in \mathbb{R} \right\},$$

$$K = \mathrm{SO}(2) = \left\{ \begin{pmatrix} \cos\theta & \sin\theta \\ -\sin\theta & \cos\theta \end{pmatrix} \right\} \quad (\theta \in [0, 2\pi]),$$

$$\mathsf{X} = \{ z \in \mathbb{C} \mid \mathcal{I}z > 0 \}.$$

The group G operates on X by the conformal transformations

$$(1) \qquad z \mapsto gz = \frac{az + b}{cz + d}, \quad g = \begin{pmatrix} a & b \\ c & d \end{pmatrix}.$$

The map $g \mapsto gi$ identifies G/K with X.

Let $\Gamma \subset G$ be a discrete subgroup of G such that $\Gamma \backslash \mathsf{X}$ has finite area. Then $\Gamma \backslash \mathsf{X}$ is either compact or the complement of finitely many points on a compact Riemann surface.

The group G, or rather $G/\{\pm 1\}$, is the group of conformal transformations of X and also the group of orientation preserving isometries of X with respect to the invariant metric

$$(2) \qquad ds^2 = (dx^2 + dy^2)/y^2$$

with volume element

$$(3) \qquad dv = dx \wedge dy/y^2$$

and curvature equal to -1.

This is one of the richest configurations of mathematics. For over a century it has been a meeting ground for analysis, number theory, and algebraic geometry, as well as the source of deep conjectures, some still open. It has been generalized to several variables in many ways, and these generalizations are the subject matter of intense activity in a broad part of mathematics.

We recall that

$$(4) \qquad \frac{dg}{dz} = (cz + d)^{-2};$$

$$(5) \qquad \mathcal{I}(gz) = \mathcal{I}(z)/|cz + d|^2.$$

We set $\mu(g, z) = cz + d$. This is an "automorphy factor", that is, $\mu(g, z)$ satisfies the cocycle identity

$$(6) \qquad \mu(gg', z) = \mu(g, g'z).\mu(g', z), \quad \text{where } g, g' \in G,\, z \in \mathsf{X}.$$

1

The most important example of Γ is the group $SL_2(\mathbb{Z})$. Its usual fundamental domain is

(7) $\{z = x + iy \in X \mid |x| \leqslant \frac{1}{2}, \ |z| \geqslant 1\}.$

0.2 Definition.   A function $f : X \to \mathbb{C}$ is called an *automorphic form* of weight m with respect to Γ if

(a) f is holomorphic and, for all $\gamma \in \Gamma$,

$$f(\gamma z) = \mu(\gamma, z)^m f(z).$$

(b) If $\Gamma \backslash X$ is not compact then some regularity conditions at the cusps are imposed.

The action of G on X extends continuously to $X \cup \mathbb{R} \cup \infty$. A point $\omega \in \mathbb{R} \cup \infty$ is *cuspidal* for Γ if it is fixed under an infinite subgroup of Γ.

Assume that ∞ is a cuspidal point for Γ. Then Γ contains a translation

$$T_c : z \mapsto z + c \quad (c \in \mathbb{R}).$$

Since $\mu(T_c, z) \equiv 1$, condition (a) shows that f is invariant under T_c. The set of real numbers r such that f is invariant under T_r is an infinite cyclic subgroup. Assume that c is a generator; hence f is periodic of period c as a function of x. The regularity condition (b) is then: f can be written as a convergent power series

$$f(z) = \sum_{n \geqslant 0} a_n e^{2\pi i n z / c}.$$

For $m = 2n$ even, an automorphic form of weight m has a geometric interpretation. Let $T^*(\Gamma \backslash X) = L$ be the cotangent bundle to $\Gamma \backslash X$, and let L^n be its nth tensor power. The automorphic forms of weight m are just the holomorphic sections of L^n or, equivalently, the holomorphic functions such that $f(z) dz^n$ is invariant under Γ.

The previous notions can be extended by allowing f to be meromorphic. In particular, for weight 0, the meromorphic automorphic forms are the meromorphic functions on $\Gamma \backslash X$ (or rather on its natural compactification; see 3.19).

0.3 (a) If $\Gamma = SL_2(\mathbb{Z})$ then the automorphic forms are called *modular* forms. We can assign to $\tau \in X$ the elliptic curve $C(\tau)$, the quotient of $\mathbb{C}$ by the lattice $\mathbb{Z} + \mathbb{Z}\tau$. Then τ and $g\tau$ ($g \in \Gamma$) define isomorphic elliptic curves, and the map $\tau \mapsto C(\tau)$ yields a bijection of $\Gamma \backslash X$ onto the set of isomorphism classes of elliptic curves. Therefore modular forms define, or are defined by, invariants of elliptic curves. In particular we have the discriminant

$$\Delta = g_2^3 - 27g_3^2 = (2\pi)^{12}q \prod_{n=1}^{\infty}(1 - q^n)^{24},$$

where

$$q = e^{2\pi i z}, \quad g_2 = 60G_2, \quad g_3 = 140G_3,$$

with G_k defined as the Eisenstein series

$$G_k = \sum_{\substack{m,n \in \mathbb{Z} \\ (m,n) \neq (0,0)}} \frac{1}{(mz + n)^{2k}}.$$

(b) *Theta series:* Let Q be a positive definite quadratic form on $\mathbb{Q}^m$ that is integral valued on $\mathbb{Z}^m$. We define a so-called theta series,

$$\theta(z, Q) = \sum_{x \in \mathbb{Z}^m} e^{2\pi i Q(x)z},$$

by its Fourier development with respect to ∞. Then

$$\theta(z, Q) = \sum_{n \geq 0} a_n(Q)e^{2\pi i n z},$$

where $a_n(Q)$ is the number of representations of n by Q, that is, the number of $x \in \mathbb{Z}^m$ such that $Q(x) = n$. It is invariant under the translations T_n ($n \in \mathbb{Z}$). To check that it is a modular form, it suffices to know that 1.1(1) is satisfied by a generating set of Γ. One such set consists of $w: z \mapsto -1/z$ and T_1. If Q is even, with determinant 1, then we have the transformation formula

$$\theta(-1/z, Q) = (-iz)^{m/2}\theta(z, Q).$$

If $m \equiv 0 \pmod 8$ then the factor $(-iz)^{m/2}$ is equal to $z^{m/2} = \mu(w, z)^{m/2}$. Hence $\theta(z, Q)$ is a modular form of weight $m/2$.

(c) The deepest relations with number theory link modular forms and Dirichlet series via the Mellin transform.

Again with respect to ∞, associate to f the Dirichlet series $D(s, f)$:

$$f(z) = \sum_{n \geq 0} a_n e^{2\pi i n z} \mapsto D(s, f) = \sum_{n \geq 1} \frac{a_n}{n^s}.$$

The Mellin transform of f is

$$(2\pi)^{-s}\Gamma(s)D(s, f) = L(s, f) = \int_0^{\infty} y^s(f(iy) - a_0)\frac{dy}{y}.$$

If f is a modular form of weight m, then

$$L(s, f) = i^m L(1 - s, f).$$

Formally, the Riemann zeta function $\zeta(2s)$ is associated in this way to the theta series $\frac{1}{2}\sum_{n \in \mathbb{Z}} e^{\pi i n^2 z}$, which (in a generalized sense) may be viewed as a modular form of weight $1/2$, for a subgroup of finite index of the modular group.

Let k be a number field. Its Dedekind zeta function is

$$\zeta_k(x) = \sum_{\mathfrak{A}} N\mathfrak{A}^{-s},$$

where $\mathfrak{A}$ rus through the integral ideals of K and $N\mathfrak{A}$ is the norm of the ideal $\mathfrak{A}$.

Hecke showed that if k is an imaginary quadratic field then there exists f (constructed as a theta series using the norm of k) such that

$$L(s, f) = \zeta_k(s).$$

Then Hecke gave as a problem to H. Maass to see whether it was also possible to define some sort of automorphic form for $SL_2(\mathbb{Z})$, or more generally for a congruence subgroup, the Mellin transform of which would yield ζ_k for k real quadratic. Recall that a subgroup Γ of $SL_2(\mathbb{Z})$ is a congruence subgroup if there exists an integer $m \geqslant 1$ such that Γ contains the kernel of the natural homomorphism $SL_2(\mathbb{Z}) \to SL_2(\mathbb{Z}/m\mathbb{Z})$ defined by reduction mod m of the coefficients.

(d) In order to solve Hecke's problem, Maass introduced some nonholomorphic automorphic forms that he called "wavefunctions". Let

$$\Delta = -y^2 \left(\frac{\partial^2}{\partial x^2} + \frac{\partial^2}{\partial y^2} \right)$$

be the Laplace–Beltrami operator on X. Consider the eigenfunctions of Δ that are Γ-invariant:

$$\Delta f = s(1 - s)f;$$

an example is

$$(\mathcal{I}z)^{(s+1)/2} \sum_{\gamma \in \Gamma_\infty \backslash \Gamma} |\mu(\gamma, x)|^{-(s+1)},$$

which converges for $\mathcal{R}s > 1$. A major problem, solved by Maass [42], was the analytic continuation of this series. For Eisenstein series in the more general case of Section 10, meromorphic continuation was proved first by Selberg [50] (see §11 for further comments).

0.4 Let

$$G \to G/K = \mathsf{X}$$

be the natural projection. G induces a map $\Gamma\backslash G \to \Gamma\backslash \mathsf{X}$. We want now to lift automorphic forms on $\Gamma\backslash \mathsf{X}$ to $\Gamma\backslash G$. To f on $\Gamma\backslash \mathsf{X}$ of weight m we associate $\tilde{f}: G \to \mathbb{C}$ given by

(1) $$\tilde{f}(g) = \mu(g, i)^{-m} f(gi).$$

The point $i \in \mathsf{X}$ is the fixed point of $K = SO(2)$. The relation 0.1(6) yields, for $g, g' \in K$,

$$\mu(gg', i) = \mu(g, i)\mu(g', i).$$

In particular,

$$k \mapsto \rho(k) = \mu(k, i)$$

is a character of K. Then $\tilde{f}$ satisfies the following conditions:

(a) $\tilde{f}(\gamma g) = \tilde{f}(g)$ for all $\gamma \in \Gamma \backslash G$;

(b) $\tilde{f}(gk) = \rho(k)^{-m} \tilde{f}(g)$;

(c) there exists a polynomial P in C (C the Casimir operator) such that

$$P(C)\tilde{f} = \left(\frac{m^2}{2} - m \right) \tilde{f};$$

(d) and a growth condition.

The *Casimir operator* C is a differential operator of second order on G that is invariant under left and right transformations (see 2.5).

The holomorphy of f implies (c). Conditions (b) and (c) are familiar in the representation theory of G; they define the so-called K-finite and $\mathcal{Z}$-finite vectors in representation spaces. Hence the study of automorphic forms is closely related to harmonic analysis on $\Gamma \backslash G$.

0.5 To summarize, we have two ways of looking at automorphic forms.

(1) As functions on X, with special behavior with respect to Γ, that may be viewed as geometric objects or functions on $\Gamma \backslash X$ (which is a Riemann surface). However, the previous examples show that automorphic forms also arise in other contexts in algebraic geometry or number theory.

(2) As objects on $\Gamma \backslash G$ that are related to the harmonic analysis on G or $\Gamma \backslash G$. This view is better suited for generalizations and the connection with L-functions (Langlands's program).

My framework will be mainly the second one, though I shall discuss some of the relations between the two. This outlook is very much influenced by the theory of infinite dimensional representations, but no knowledge of this is required until Section 14.

The systematic use of convolution with smooth functions with compact support and in particular of 2.14 is a technique I learned from Harish-Chandra (orally and in an unpublished manuscript of his proving the results announced in [29]). One goal of this exposition is to facilitate access to the general case of a reductive group. Many of the proofs given here are merely adaptations of arguments in more general situations (see notably [2; 4; 21; 22; 31; 41]). I have also tried to use notation suggestive of the general case.

0.6 We now summarize the contents of the various sections. As before,

$$G = \mathrm{SL}_2(\mathbb{R}), \qquad K = \mathrm{SO}(2),$$

and $\mathsf{X} = G/K$ is the upper half-plane.

Section 1 lists the prerequisites and some general notation to be used without further reference. Section 2 reviews or introduces the basic notions and facts concerning $\mathrm{SL}_2(\mathbb{R})$ and invariant differential operators on $\mathrm{SL}_2(\mathbb{R})$ or X. As remarked previously, 2.14 is all-important, but the technique of the proof will not play a role in the sequel. Section 3 is mainly devoted to reduction theory – that is, the construction of a fundamental domain for the action on X (or G) of a discrete subgroup Γ of G. We first use Poincaré's method, then prove Siegel's theorem to the effect that the Poincaré fundamental domain has finitely many sides if $\Gamma \backslash \mathsf{X}$ has finite area, and finally describe a somewhat rougher fundamental domain in terms of Siegel sets. This is the form to be used in the sequel. A last section gives a number of examples of such Γ. Section 4 describes the unit disc model of X, which is sometimes more convenient than the upper half-plane. Section 5 introduces the notion of automorphic form, proves some first properties, and relates to it the classical notion of holomorphic automorphic form of a given weight. Section 6 is devoted to Poincaré series, or rather to a generalization of the series introduced by Poincaré. Section 7 introduces the so-called constant term at a cusp (it is indeed a constant in the classical holomorphic case, but a function in general) and then proves an estimate for the difference between an automorphic form and its constant term at a cusp in the neighborhood of that cusp. This estimate is used in Section 8 to prove the finite dimensionality of the space of automorphic forms of a given type and in Section 9 to show that convolution by a smooth compactly supported function on G is a Hilbert–Schmidt operator on the space of cuspidal functions (which is a closed G-invariant subspace of $L^2(\Gamma \backslash G)$), and also on a somewhat bigger subspace.

In Section 8 it is also shown that a Poincaré series is a cusp form. Section 10 introduces Eisenstein series, which are automorphic forms orthogonal to cusp forms. The definition depends on the choice of a cuspidal parabolic subgroup P, a complex parameter s, and a character of $\mathrm{SO}(2)$. These series are shown to converge and represent an automorphic form if $\mathcal{R}s > 1$. Section 11 is devoted to their analytic continuation as a function of s. In this section the existence of a meromorphic continuation satisfying a functional equation is proved by a method, learned from J. Bernstein, that is an abstraction of A. Selberg's "third proof". More information on the poles in the right half-plane $\mathcal{R}s \geqslant 0$ is obtained in Section 12. For $\mathcal{R}s > 0$, it is deduced from the Maass–Selberg relations, which give the scalar product of two truncated Eisenstein series (12.10). These relations are extended to the case where one argument is an automorphic form (12.16, 12.22) and used to discuss the space of automorphic forms orthogonal to cusp forms and its relations with Eisenstein series (12.17, 12.24). These results are applied in Section 13 to describe the spectral decomposition with respect to the Casimir operator (which is shown to be essentially self-adjoint) of the part of $L^2(\Gamma\backslash G)$ on which K acts via a given character.

Section 14 recalls a number of definitions and basic facts about infinite dimensional representations of G. We stick in principle to $\mathrm{SL}_2(\mathbb{R})$, but it is rather clear that the natural framework is much more general. Section 15 describes various representations of $\mathrm{SL}_2(\mathbb{R})$: principal series, their decomposition when they are not irreducible, enumeration of the irreducible unitary representations, holomorphic or antiholomorphic realizations of discrete series representations, and direct integrals of unitary principal series. Sections 16 and 17 are devoted to the spectral decomposition of $L^2(\Gamma\backslash G)$. In Section 16 it is shown (16.2, 16.6) that the discrete spectrum is the direct sum of the closure of the space of cusp forms, of finitely many complementary principal series spanned by residues of Eisenstein series at poles on $(0, 1)$, and of the constants (which are residues of Eisenstein series at 1). In Section 17 it is shown that the continuous spectrum is a finite orthogonal sum of direct integrals of unitary principal series, parameterized by the cusps (17.7). Finally, Section 18 provides some brief indications on further developments and literature.

Basic material on $SL_2(\mathbb{R})$, discrete subgroups, and the upper half-plane

1 Prerequisites and notation

Our main prerequisites may be listed as follows.

1.1 *Elementary theory of Lie groups and Lie algebras*, and the interpretation of the elements of the Lie algebra as differential operators on the group or its coset spaces. This will be used only for $SL_2(\mathbb{R})$, its Lie subgroups, and the upper half-plane. The book by Warner [58] is more than sufficient for our needs, except for some facts on Haar measures (2.9, 10.9).

1.2 *The regularity theorem for elliptic operators* (see the remark in 2.13 for references).

1.3 *Some functional analysis*, mainly about operators on Hilbert spaces. For the sake of definiteness, I have used two basic textbooks ([46] and [51]) and have given at least one precise reference for every theorem used. But this material is standard and the reader is likely to find what is needed in his (or her) favorite book on functional analysis. The demands will increase as we go along, and the material will be briefly reviewed before it is needed. Such review is mainly intended to refresh memory, fix notation, and give references, not to be a full-fledged introduction.

1.4 *Infinite dimensional representations of G*. We shall review what we need in Sections 14 and 15 and also give a few proofs, but mostly refer to the literature. An essentially self-contained discussion of some of the results on $SL_2(\mathbb{R})$ stated in Sections 2 and 14 is contained in the first part of [40].

The following notation will be used without further ado.

1.5 $\mathbb{Z}$ is the ring of integers, $\mathbb{N} = \{z \in \mathbb{Z}, z \geqslant 0\}$ the set of natural numbers, $\mathbb{Q}$ (resp. $\mathbb{R}$, $\mathbb{C}$) the field of rational (resp. real, complex) numbers, $\mathbb{R}^+$ the additive group of positive real numbers, and $\mathbb{R}^{*+}$ the multiplicative group of strictly positive real numbers.

1.6 Let f, g be strictly positive functions defined on some set X. We write $f \prec g$ if there exists $c > 0$ such that $f(x) \leqslant cg(x)$ for all $x \in X$; we write $f \succ g$ if $g \prec f$,

9

and $f \asymp g$ if $f \prec g$ and $g \prec f$. Thus $f \asymp g$ if and only there exists $c > 0$ such that

$$c^{-1} f(x) \leqslant g(x) \leqslant c f(x) \quad (x \in X).$$

If $f \prec g$ then we say that f is *essentially bounded* by g.

1.7   If X is a topological space then $C(X)$ denotes the $\mathbb{C}$-algebra of complex valued continuous functions on X. If X is locally compact, then $C_c(X)$ is the space of $f \in C(X)$ with compact support.

1.8   Let G be a group. Its center is denoted CG. If $g \in G$, then Int g is the inner automorphism $x \mapsto g.x.g^{-1}$ $(x \in G)$. We also write $^g x$ for $g.x.g^{-1}$. More generally, if $E \subset G$, then $^g E = \{ g.x.g^{-1} \mid x \in E \}$. If E is a subgroup, its normalizer $\mathcal{N}_G E$ (resp., its centralizer $\mathcal{Z}_G E$) is

(1)
$$\mathcal{N}_G E = \{ g \in G \mid \text{Int } g(E) = E \},$$
$$\mathcal{Z}_G(E) = \{ g \in G \mid \text{Int } g(x) = x, \ x \in E \}.$$

If $D \subset G$, then $D^{-1} = \{ x^{-1}, x \in D \}$ and D is symmetric if $D = D^{-1}$; that is, D is stable under the "inversion" $x \mapsto x^{-1}$.

If G operates on the space X, then G_x denotes the isotropy group at $x \in X$: $G_x = \{ g \in G, \ g \cdot x = x \}$.

1.9   If G is a Lie group, then $\text{Ad } x$ $(x \in G)$ is the differential at the identity of Int g. It is an automorphism of the Lie algebra of G. In general, the Lie algebra of a Lie group $G, P, N, \ldots$ is denoted by the corresponding German lower-case letter $\mathfrak{g}, \mathfrak{p}, \mathfrak{n}, \ldots$. We also write $^x X$ for $\text{Ad } x(X)$ $(x \in G, X \in \mathfrak{g})$.

1.10   If R is a commutative ring, then $\text{SL}_2(R)$ is the group of 2×2 matrices with coefficients in R and determinant 1.

SO(2) is the special orthogonal group on $\mathbb{R}^2$:

$$\text{SO}(2) = \left\{ \begin{pmatrix} \cos \varphi & \sin \varphi \\ -\sin \varphi & \cos \varphi \end{pmatrix} \middle| \ \varphi \in [0, 2\pi] \right\};$$
$$X = \{ z \in \mathbb{C} \mid \mathcal{I}z > 0 \},$$

the upper half-plane.

1.11   If α is a complex valued function on a group G, then $\check{\alpha}$ (resp. α^*) is the function

$$g \mapsto \alpha(g^{-1}) \quad \left(\text{resp. } g \mapsto \overline{\alpha(g^{-1})} \right).$$

1.12 Let M be a smooth manifold. Then $C^\infty(M)$ is the algebra of smooth complex valued functions on M and, given $k \in \mathbb{N}$, $C^k(M)$ is the algebra of complex valued functions which, together with their derivatives up to order k, are continuous.

2 Review of $SL_2(\mathbb{R})$, differential operators, and convolution

2.1 In the sequel, G stands for $SL_2(\mathbb{R})$ and PG for $SL_2(\mathbb{R})/(\pm Id)$. For $g \in G$, we let l_g and r_g be the left and right translations. On functions, we have

$$l_g f(x) = f(g^{-1}x), \qquad r_g f(x) = f(xg).$$

In both cases, G operates on the left; that is,

$$l_{gg'} = l_g l_{g'}, \quad r_{gg'} = r_g r_{g'} \quad (g, g' \in G).$$

The Lie algebra $\mathrm{Lie}(G)$ or $\mathfrak{g}$ of G is

$$\mathrm{Lie}(G) = \mathfrak{g} = \{ M \in M_2(\mathbb{R}) \mid \mathrm{trace}\, M = 0 \}.$$

To every $Y \in \mathfrak{g}$ is associated the one-parameter subgroup:

$$t \mapsto e^{tY} = \sum_{n=0}^{\infty} \frac{t^n Y^n}{n!} \quad (t \in \mathbb{R}).$$

The standard basis of $\mathfrak{g}$ consists of H, E, F given by

(1) $$H = \begin{pmatrix} 1 & 0 \\ 0 & -1 \end{pmatrix}, \quad E = \begin{pmatrix} 0 & 1 \\ 0 & 0 \end{pmatrix}, \quad F = \begin{pmatrix} 0 & 0 \\ 1 & 0 \end{pmatrix}.$$

They satisfy the commutator relations

(2) $$[H, E] = 2E, \quad [H, F] = -2F, \quad [E, F] = H.$$

The group G operates on itself by inner automorphisms and on $\mathfrak{g}$ by the adjoint representation. We let $\mathrm{Int}\, g$ denote the automorphism $x \mapsto g.x.g^{-1}$ of G and $\mathrm{Ad}\, g : \mathfrak{g} \to \mathfrak{g}$ its differential at the identity element. In our case we have $\mathrm{Ad}\, g(X) = g.X.g^{-1}$ (matrix product). Note the relation

(3) $$g.e^{tX}.g^{-1} = e^{t\,\mathrm{Ad}\,g(X)} \quad (t \in \mathbb{R},\ g \in G,\ X \in \mathfrak{g}).$$

We shall identify $Y \in \mathfrak{g}$ with the left-invariant differential operator on G defined by

(4) $$Yf(x) = \frac{d}{dt} f(xe^{tY})\Big|_{t=0} \quad (f \in C^\infty(G),\ x \in G,\ Y \in \mathfrak{g}).$$

Occasionally, we shall also consider the right-invariant differential operator Y_r associated to $Y \in \mathfrak{g}$ by the rule

(5) $$Y_r f(x) = \frac{d}{dt} f(e^{-tY}.x)\Big|_{t=0} \quad (f \in C^\infty(G),\ x \in G,\ Y \in \mathfrak{g}).$$

12

The Y (resp. Y_r) generate over $\mathbb{C}$ the algebra of left-invariant (resp. right-invariant) differential operators on G. It is isomorphic (resp. anti-isomorphic) to the universal enveloping algebra $\mathcal{U}(\mathfrak{g})$ of $\mathfrak{g}$ over $\mathbb{C}$. We identify $\mathcal{U}(\mathfrak{g})$ with the algebra of left-invariant differential operators. The right-invariant differential operator associated to $D \in \mathcal{U}(\mathfrak{g})$ will be denoted D_r.

Because the Y ($Y \in \mathfrak{g}$) span the tangent space at every point, the first partial derivatives with respect to local coordinates in a chart are linear combinations, with coefficients in $C^\infty(G)$, of the Y_r; hence the higher partial derivatives are linear combinations with coefficients in $C^\infty(G)$ of the D_r ($D \in \mathcal{U}(\mathfrak{g})$). As a consequence, if $f \in C^\infty(G)$ and $x \in G$ are such that $D_r f(x) = 0$ for all $D \in \mathcal{U}(\mathfrak{g})$, then all the partial derivatives of f vanish at x.

By the Poincaré–Birkhoff–Witt theorem (cf. e.g. [12, §2, no. 7]), for any basis A, B, C of $\mathfrak{g}$, the monomials $A^m.B^n.C^p$ ($m, n, p \in \mathbb{N}$) form a vector space basis of $\mathcal{U}(\mathfrak{g})$.

Any left-invariant differential operator is a finite linear combination of compositions $X_1 \ldots X_s$ ($X_i \in \mathfrak{g}$). The action of such an operator on f is given by

$$(6) \qquad X_1 \ldots X_s f(x) = \left. \frac{d^s}{dt_1 \ldots dt_s} f(xe^{t_1 X_1} \ldots e^{t_s X_s}) \right|_{t_1 = t_2 = \cdots = t_s = 0}.$$

2.2 Convolution. Let α, β be two locally integrable functions on G. The convolution $\alpha * \beta$ is defined by

$$(\alpha * \beta)(x) = \int_G \alpha(xy)\beta(y^{-1}) \, dy = \int_G \alpha(y)\beta(y^{-1}x) \, dy,$$

assuming the integral exists. This is the case – the only one we will consider – if, for instance, at least one of them has compact support; the convolution extends to distributions. If we identify $\mathcal{U}(\mathfrak{g})$ with the distributions supported by the origin, we can write

$$(1) \qquad\qquad Yf(x) = (f * (-Y))(x) \quad (Y \in \mathfrak{g}).$$

In terms of convolution, the effect of the right-invariant differential operator Y_r defined by Y on f can be written

$$Y * f \quad (y \in \mathfrak{g}).$$

Thus we can also write $*Y$ for $-Y$ and $Y*$ for Y_r.

The reader unfamiliar with distribution theory may simply view this as defining $f * (-Y)$ and $Y * f$. This notation will be convenient on the few occasions where we consider simultaneously the left-invariant and right-invariant differential operators defined by $Y \in \mathfrak{g}$.

2.3   (i) *Convolution is a smoothing operator*: If α is continuous and $\beta \in C_c^\infty(G)$, then $\alpha * \beta \in C^\infty(G)$. In fact, if $D \in \mathcal{U}(\mathfrak{g})$ then

(1) $$D(\alpha * \beta) = \alpha * (D\beta),$$

as follows from the definition.

     (ii) *Convolution is associative*: If f, α, β are locally L^1 and α, β have compact supports, then

(2) $$(f * \alpha) * \beta = f * (\alpha * \beta).$$

This is a straightforward computation that is left to the reader. (Note that the double integral defining the left-hand side is absolutely convergent, so that we may change the order of integration.)

     (iii) If $\alpha \in C^\infty(G)$, $\beta \in C_c^\infty(G)$, and $D \in \mathcal{U}(\mathfrak{g})$ is *bi-invariant* (i.e., belongs to $\mathcal{Z}(\mathfrak{g})$; cf. 2.5), then

(3) $$D(\alpha * \beta) = (D\alpha) * \beta.$$

Proof. By assumption, D commutes with right translations. Therefore

$$((D\alpha) * \beta)(x) = \int_G r_y D\alpha(x).\beta(y^{-1})\,dy = \int_G (Dr_y\alpha)(x).\beta(y^{-1})\,dy.$$

The integral is in fact over any compact set containing the support of $\check{\beta}$, hence D and $\int_G$ can be permuted to give

$$((D\alpha) * \beta)(x) = D\left(\int_G r_y\alpha(x).\beta(y^{-1})\,dy \right) = D(\alpha * \beta)(x).$$

2.4 Dirac sequences.   A sequence of functions $\alpha_n \in C_c^\infty(G)$ ($n \in \mathbb{N}$) is called a *Dirac sequence* if $\alpha_n(x) \geqslant 0$ for all $x \in G$, $n \in \mathbb{N}$ and

(1) $$\text{support}\,(\alpha_n)_{n\to\infty} \to \{1\},$$

(2) $$\int_G \alpha_n(x)\,dx = 1 \quad (n \in \mathbb{N}).$$

Then, for any $f \in C(G)$, the sequence $f * \alpha_n$ converges to f; the convergence is absolute and uniform on compact sets for f and for each derivative Df ($D \in U(\mathfrak{g})$).

Proof. We must show that given $D \in \mathcal{U}(\mathfrak{g})$, a compact set C in G, and $\varepsilon > 0$, there exists j such that

(3) $$|(Df * \alpha_j)(x) - Df(x)| \leqslant \varepsilon \quad (x \in C).$$

There exists an open relatively compact neighborhood U of 1, which we may assume to be symmetric, such that

(4) $|Df(x.y) - Df(x)| \leqslant \varepsilon \quad (x \in C, \ y \in U)$.

Then, in view of (2),

$$(Df * \alpha_j)(x) - f(x) = \int_G Df(x.y^{-1})\alpha_j(y)\,dy - \int_G f(x)\alpha_j(y)\,dy$$

$$|(Df * \alpha_j)(x) - Df(x)| \leqslant \int_G |Df(x.y^{-1}) - Df(x)|\alpha_j(y)\,dy.$$

We may find j such that U contains the support $\operatorname{supp}\alpha_j$ of α_j. Then the integral on G is in fact on U and so, taking (4) into account, we have

$$|(Df * \alpha_j)(x) - Df(x)| \leqslant \int_U |Df(x.y^{-1}) - Df(x)|\alpha_j(y)\,dy$$

$$\leqslant \varepsilon \int_U \alpha_j(y)\,dy = \varepsilon$$

as was to be proved.

2.5 Bi-invariant differential operators. The left-invariant differential operators that are also right-invariant form the center $\mathcal{Z}(\mathfrak{g})$ of $\mathcal{U}(\mathfrak{g})$. The algebra $\mathcal{Z}(\mathfrak{g})$ is a polynomial algebra $\mathbb{C}[C]$ in the Casimir operator. In the standard basis (2.1), we have

(1) $C = \frac{1}{2}H^2 + EF + FE = \frac{1}{2}H^2 + H + 2FE = \frac{1}{2}H^2 - H + 2EF$.

The invariant definition of a Casimir operator is as follows. Let $K(X, Y)$ be the Killing form; that is, let

(2) $K(X, Y) = \operatorname{tr}(\operatorname{ad} X \circ \operatorname{ad} Y) \quad (X, Y \in \mathcal{G})$.

Then, if X_1, X_2, X_3 is a basis of $\mathfrak{g}$ and X_1', X_2', X_3' is the dual basis with respect to K, a Casimir operator is given by

$$C = \sum_{i=1}^{3} X_i X_i'.$$

We have used the Killing form, but the trace form $T(X, Y) = \operatorname{tr} XY$ in any irreducible representation is a nonzero constant multiple of the Killing form and would also be suitable. For instance, the trace of the identity representation in $\mathbb{R}^2$ is readily checked to be equal to one fourth of the Killing form.

The Lie algebra also defines differential operators on any homogeneous space or, more generally, on any smooth manifold on which G acts smoothly. Consider in particular the upper half-plane X and denote by V^* the vector field on X defined by $V \in \mathfrak{g}$. Since G operates on the left on X $= G/K$, the map $V \mapsto V^*$ extends to an anti-isomorphism of $\mathcal{U}(\mathfrak{g})$ onto an algebra of differential operators on X. Write

$$e^{tV}(z) = x(t) + i.y(t) \quad (z \in X).$$

Then the value V_z^* of V^* at $z \in X$ is

$$\frac{dx(t)}{dt}\bigg|_{t=0} \frac{\partial}{\partial x} + \frac{dy(t)}{dt}\bigg|_{t=0} \frac{\partial}{\partial y}.$$

From this, it is easy to check that

$$E^* = \frac{\partial}{\partial x},$$

$$F^* = (y^2 - x^2)\frac{\partial}{\partial x} - 2xy\frac{\partial}{\partial y},$$

(3)

$$H^* = 2\left(x\frac{\partial}{\partial x} + y\frac{\partial}{\partial y}\right),$$

$$C^* = 2y^2\left(\frac{\partial^2}{\partial x^2} + \frac{\partial^2}{\partial y^2}\right).$$

2.6 Parabolic subgroups.   We let

$$N_0 = \left\{\begin{pmatrix} 1 & x \\ 0 & 1 \end{pmatrix} \bigg| x \in \mathbb{R}\right\}, \quad A_0 = \left\{\begin{pmatrix} t & 0 \\ 0 & t^{-1} \end{pmatrix} \bigg| t > 0\right\}, \quad K = \mathrm{SO}(2),$$

and

$$P_0 = \left\{\begin{pmatrix} a & b \\ c & d \end{pmatrix} \in \mathrm{SL}_2(\mathbb{R}) \bigg| c = 0\right\}.$$

Then $P_0 = \pm P_0^\circ$, where $P_0^\circ = N_0 A_0$, that is, $P_0 = CG.P_0^\circ$. The group P_0 is the stability group in G of the line $\mathbb{R}.e_1$, and N_0 is the derived group of P_0 or of $N_0 A_0$. The group P_0° is the semidirect product of N_0 and any conjugate A of A_0, to be referred to as a *Cartan subgroup* of P_0°.

A parabolic subgroup P of G is, by definition here, the stability group of some line in $\mathbb{R}^2$. Since K is transitive on those subgroups, P is conjugate to P_0 under K. The properties of P_0 just listed carry over to P. In particular, $P^\circ = N_P.A$, where N_P is the derived group of P or P° and A is any conjugate of A_0 in P, also to be called a Cartan subgroup.

It is clear that A_0 is the unique Cartan subgroup of P_0 with Lie algebra orthogonal to that of K. Therefore P has a unique Cartan subgroup with that same property. A p-pair (P, A) consists of a parabolic subgroup and a Cartan subgroup. If A is the unique Cartan subgroup of P with Lie algebra orthogonal to that of K, then the p-pair is said to be *normal*. The group K is transitive on the normal p-pairs, and K/CG is simply transitive on them. In the sequel, a p-pair is assumed to be normal, unless otherwise stated.

2.7 Iwasawa and Bruhat decompositions.   Let (P, A) be a p-pair, and let $N = N_P$. Then we have the Iwasawa decomposition

(1) $$G = NAK,$$

with uniqueness of decomposition. That is, we have uniquely

$$g = n(g)a(g)k(g) \quad (n(g) \in N,\ a(g) \in A,\ k(g) \in K),$$

and $g \mapsto (n(g), a(g), k(g))$ is a diffeomorphism of G onto $N \times A \times K$.

Let $\mathcal{N}(A)$ be the normalizer of A. Then the "Weyl group"

$$W(A, G) := \mathcal{N}(A)/\mathcal{Z}_G(A) = \{1, w\}$$

is of order 2 and w acts on A by the inversion $a \mapsto a^{-1}$. As a special case of the Bruhat decomposition we have

(2) $$G = P.W(A, G).N = P \cup PwN$$

(disjoint union). Using conjugation by K, we see that it suffices to prove this (or the Iwasawa decomposition) for $P = P_0$. Then w is represented by the matrix $\begin{pmatrix} 0 & 1 \\ -1 & 0 \end{pmatrix}$, and (1) means that if $g = \begin{pmatrix} a & b \\ c & d \end{pmatrix}$ is such that $c \neq 0$ then it can be written (uniquely) as a product $b.w.n$ ($b \in P_0$, $n \in N_0$); this follows from a straightforward computation.

2.8 We let $X(A)$ be the group of characters of A, that is, of continuous homomorphisms $A \mapsto \mathbb{C}^*$; it is isomorphic to $\mathbb{C}$. We want to fix such an isomorphism by a rule valid for any p-pair (P, A). Let D_P be the line on $\mathbb{R}^2$ stable under P and let e_P be a unit vector in D_P. Then we define $\rho \in X(A)$ by the rule

$$a^\rho = \|a.e_P\|.$$

Here $\|\cdot\|$ is the Euclidean norm, which does not change if we replace e_P by $-e_P$. For instance, if $(P, A) = (P_0, A_0)$ is the standard p-pair then $D_P = \mathbb{R}.e_1$, and if we write

$$a = \begin{pmatrix} t_a & 0 \\ 0 & t_a^{-1} \end{pmatrix} \quad (t_a \in \mathbb{R},\ t_a > 0)$$

then $a^\rho = t_a$. In the sequel, we identify $\mathbb{C}$ with $X(A)$ by assigning to $s \in \mathbb{C}$ the character

$$a \mapsto a^{s\rho} = \|a.e_P\|^s.$$

This map identifies $\mathbb{R}$ with the group of real valued characters of A, to be denoted $X(A)_{\mathbb{R}}$. Note that $\pm A$ is the intersection of two parabolic subgroups, say P and P^-. If we view A as a Cartan subgroup of P^- then a^ρ is replaced by $a^{-\rho}$.

Writing

(1) $$x = na(x)k \quad (n \in N,\ a(x) \in A,\ k \in K),$$

we have

(2) $$\|x^{-1}e_P\| = \|k^{-1}a(x)^{-1}n^{-1}e_P\| = a(x)^{-\rho},$$

since k leaves $\|\cdot\|$ invariant and $ne_P = e_P$. We set

(3) $$h_P(x) = \|x^{-1}e_P\|^{-1};$$

hence

$$h_P(x) = a(x)^{\rho}.$$

Note that

(4) $$h_P(naxk) = h_P(a).h_P(x) \quad (n \in N, a \in A, k \in K, x \in G),$$

which follows immediately from the definition, and

(5) $$h_{P'}(^kx) = h_P(x) \quad (x \in G, k \in K, P' = {}^kP).$$

This also follows from the definitions once it is noted that $k.e_P = \pm e_{P'}$. The function $h_P(x)$ will play an important role in the discussion of Eisenstein series.

Let Y be in the Lie algebra of N. Then

(6) $$\operatorname{Ad} a(Y) = a.Y.a^{-1} = a^{2\rho}.Y \quad (a \in A).$$

By conjugation, it suffices to check this in the standard case $P = P_0$. Then, with $Y = \begin{pmatrix} 0 & c \\ 0 & 0 \end{pmatrix}$ we have

(7) $$a.Y.a^{-1} = \begin{pmatrix} t_a & 0 \\ 0 & t_a^{-1} \end{pmatrix} \begin{pmatrix} 0 & c \\ 0 & 0 \end{pmatrix} \begin{pmatrix} t_a^{-1} & 0 \\ 0 & t_a \end{pmatrix} = t_a^2 \begin{pmatrix} 0 & c \\ 0 & 0 \end{pmatrix} = a^{2\rho}Y.$$

We shall also view 2ρ as a character of P that is trivial on N and CG. If N is identified with its Lie algebra by the exponential map, we have

(8) $$x.n.x^{-1} = h_P(x)^{2\rho}.n \quad (x \in P, n \in N).$$

Note that in the standard case $P = P_0$ we have

(9) $$h_P(x) = \|x^{-1}.e_1\|^{-1} = (c^2 + d^2)^{-1/2}, \quad \left(x = \begin{pmatrix} a & b \\ c & d \end{pmatrix}\right).$$

2.9 Haar measures. We recall a few general facts to set the framework, but they will be used only for G, P, N, A, and some quotients by closed subgroups.

(a) A positive (nonzero) measure dh on a locally compact group H is a left- (resp. right-) invariant Haar measure if a measurable set and its left (resp. right) translates have the same measure, or, equivalently, if for all $g \in H$ and $f \in C_c(H)$ the function $l_g f$ (resp. $r_g f$) have the same integral on H with respect to dh. There is always one such measure, unique up to a strictly positive factor. Given a left-invariant Haar measure dh, there is a homomorphism Δ or $\Delta_H : H \mapsto R^{*+}$, the *modulus* of H, such that $r_g(dh) = \Delta(h)dh$. Then $\Delta(h)dh$ is right-invariant: Let $g \in H$, $h' = h.g$, and $f \in C_c(H)$; then

(1) $$\int_H f(hg)\Delta(h)\,dh = \int_H f(h')\Delta(h')\Delta(g)^{-1}r_g(dh) = \int_H f(h')\,dk'.$$

H is *unimodular* if a left-invariant measure is right-invariant, that is, if $\Delta \equiv 1$.

(b) Assume H and a closed subgroup L to be unimodular, and let dh, dl be Haar measures on H and L respectively. Then there is a unique H-invariant quotient measure $d\dot{h}$ on H/L such that $dh = d\dot{h}.dl$, meaning that for $f \in C_c(G)$ we have

(2) $$\int_H f(h)\,dh = \int_{H/L} d\dot{h} \int_L f(h.l)\,dl.$$

This then extends to measurable functions by standard integration theory. In particular, if f is continuous and integrable on H, then $\dot{h} \mapsto \int_L f(l.h)\,dl$ is integrable on $L\backslash H$ and (2) holds. Conversely, if f is such that $\dot{h} \mapsto \int_L |f(l.h)|\,dl$ is integrable on $L\backslash H$, then f is integrable on H.

There are of course similar statements with "left" and "right" interchanged. The modulus for going from left invariance to right invariance is the inverse of the previous Δ_H.

(c) Let now H be a Lie group, and let $\omega_1, \ldots, \omega_n$ ($n = \dim H$) be a basis of the space of the left- (resp. right-) invariant 1-forms. Then $\omega_1 \wedge \cdots \wedge \omega_n$ defines a left- (resp. right-) invariant Haar measure and $\Delta(h) = |\det \mathrm{Ad}\, h|$. This modulus is clearly trivial for G, K, N, A. We let dg, dk, dn, da denote Haar measures on these groups; dk is normalized so that K has measure 1. On N identified with $\mathbb{R}$ as before, dn is the usual Lebesgue measure. On A, take as coordinate $t = a^\rho$; then $da = dt/t$. On P° the measure $dp = dn.da$ is right-invariant:

$$r_{q.b}(dn\,da) = r_{b.n.b^{-1}}(dn).r_b da = dn.da \quad (q \in N, \; b \in A),$$

but $\det \mathrm{Ad}\, q.b = b^{2\rho}$, so P° is not unimodular. The measure

(3) $$d_l p = a^{-2\rho}.dn.da = h_P(a)^{-2}.dn.da$$

is left-invariant. The quotient G/K is diffeomorphic to P°, and the G-invariant measure is the image of $d_l p$. On G we may (and do) take as Haar measure

(4) $$d_l p.dk = h_P(a)^{-2}.dn.da.dk.$$

An essentially self-contained discussion of what we need may be found in [40, III, §1]. For a general treatment of Haar measures, see [10].

We also have $G/K = \mathsf{X}$, and $dx \wedge dy.y^{-2}$ is a G-invariant measure. So we can also take $y^{-2}.dx \wedge dy.dk$ as Haar measure on G. It can be shown that this is twice the measure given by (4) (see [40]).

2.10　Cartan decomposition.　Let $\mathfrak{s}_o$ be the space of symmetric 2×2 real matrices of trace 0, and let $S_o = \exp \mathfrak{s}_o$. Then S_o is the space of positive nondegenerate symmetric matrices of determinant 1 and we have the "polar decomposition"

$$G = S_o K.$$

Given $g \in G$, it follows that

$$g = sk \quad (s \in S_o, k \in K) \quad \text{with } s = (g^t g)^{1/2}.$$

This is a special case of the Cartan decomposition of a semisimple group. The Lie algebra $\mathfrak{k}$ of K is spanned by

$$W = \begin{pmatrix} 0 & 1 \\ -1 & 0 \end{pmatrix} = E - F,$$

and $\mathfrak{s}_0$ is spanned by H and by

$$Z = \begin{pmatrix} 0 & 1 \\ 1 & 0 \end{pmatrix} = E + F.$$

In this basis,

(1) $$C = \tfrac{1}{2}(H^2 + Z^2 - W^2);$$

therefore

(2) $$\Omega = C + W^2 = \tfrac{1}{2}(H^2 + Z^2 + W^2)$$

is an elliptic operator.

2.11　Characters of K.　The group K is the group of rotations of $\mathbb{R}^2$ around the origin and may be identified with the group $\{e^{i\varphi}\}$ of complex numbers of modulus 1. Let $\varphi \mapsto k_\varphi$ be an isomorphism of $\mathbb{R}/2\pi$ onto K. The characters of K are its 1-dimensional representations. They form a group $\hat{K}$ isomorphic to $\mathbb{Z}$, where the character

$$\chi_m : k_\varphi \mapsto e^{im\varphi}$$

is assigned to $m \in \mathbb{Z}$. Any continuous finite dimensional representation of K is a direct sum of 1-dimensional ones. We shall say that a function f on G is of right (resp. left) K-type m if for all $x \in G$ and $k \in K$ we have

(1) $$f(x.k) = \chi_m(k)f(x) \quad (\text{resp. } f(k^{-1}.x) = \chi_m(k)f(x)).$$

2.12　K-finite and $\mathcal{Z}$-finite functions.　A function f on G is K-finite on the right (resp. left) if the right (resp. left) translations $r_k f$ (resp. $l_k f$) span a finite dimensional vector space ($k \in K$). Any K-finite function f is a finite sum $f = \sum f_i$, where f_i is of some type m_i.

The function f is $\mathcal{Z}$-finite or $\mathcal{C}$-finite if the $\mathcal{C}^n f$ ($n \in \mathbb{N}$, $\mathcal{C}$ the Casimir operator) are contained in a finite dimensional vector space of functions. This is equivalent to either of the following conditions:

(i) there exists a nonconstant polynomial P in one variable such that $P(\mathcal{C})f = 0$;
(ii) there exists an ideal I of finite codimension of $\mathbb{C}[\mathcal{C}]$ such that $Df = 0$ for any $D \in I$.

In fact, $I = (P(\mathcal{C}))$ is the ideal generated by $P(\mathcal{C})$.

In general, f will be smooth. Of course, the definition of K-finiteness makes sense for any function. The notion of $\mathcal{Z}$-finiteness extends to any locally integrable function, if meant in the sense of distributions. We have

(1) $$Wf = imf \quad \text{if } f \text{ is of right } K\text{-type } m.$$

In fact,

$$Wf(x) = \frac{d}{dt}(f(xe^{tW}))\Big|_{t=0} = \frac{d}{dt}(f(x)e^{imt})\Big|_{t=0} = imf(x).$$

There is a converse. Remembering that

$$k_\varphi = \exp\varphi.W = \sum_{n \geq 0} \frac{\varphi^n}{n!} W^n,$$

we see that if $Wf = imf$ then

$$r_k f(x) = e^{mi\varphi} f(x)$$

and hence $r_k f = \chi_m(k) f$, that is, f is of type m.

In the sequel, we let

(2) $$I_c^\infty(G) = \{\alpha \in C_c^\infty(G) \mid \alpha(xk) = \alpha(kx), k \in K, x \in G\}.$$

Then, if f is of K-type m on the right, so is $f * \alpha$ for $\alpha \in I_c^\infty(G)$. The functions on G satisfying the condition in (2) are called *K-invariant*.

2.13 Theorem. *Let f be $\mathcal{Z}$-finite and K-finite on one side. Then f is smooth.*

Proof. This will follow from the regularity theorem for elliptic operators. We carry out the reduction assuming f to be K-finite on the right. We may assume it is of type m for some $m \in \mathbb{N}$. By assumption there exists a nonconstant polynomial P in one variable such that

$$P(\mathcal{C})f = 0.$$

We may assume P to be monic. Let n be its degree. Then

$$P(\mathcal{C}) = \prod_{i=1}^{n}(\mathcal{C} - \lambda_i).$$

On the other hand,

$$Wf = imf,$$

so

$$W^2 f = -m^2 f.$$

Then

$$P(\mathcal{C})f = 0 \iff \prod_{i=1}^{n}(\mathcal{C} + W^2 - \lambda_i + m^2)f = 0,$$

that is,

$$Q(\Omega)f = 0,$$

where Q is a monic nonconstant polynomial and Ω is defined by 2.10(2). The operator $Q(\Omega)$ is elliptic, so f is smooth by the regularity theorem.

Remark. The regularity theorem is classical and may be found in many places. For a self-contained treatment in the smooth case and other references, see [40, Apx. 4] or [58, pp. 227–49].

     The theorem is stronger and is valid for a much more general class of functions – in fact, for distributions. Note first that the notion of K-finiteness makes sense for any function on G. If f is a local L^1-function (or a distribution) then $\mathcal{C}^n f$ in the distribution sense is the distribution

$$\varphi \mapsto \int_G f(x)\mathcal{C}^n\varphi(x)\,dx \quad (\varphi \in C_c^\infty(G),\ n \in \mathbb{N});$$

it is $\mathcal{C}$-finite if those distributions for $n \in \mathbb{N}$ span over $\mathbb{C}$ a finite dimensional vector space of distributions. The strong form of the theorem [35, Thm. 7.41] implies that if f is locally L^1, K-finite on one side, and $\mathcal{Z}$-finite, then it is smooth. In our case, the manifold and differential operators are analytic, and then it can be proved that f is analytic [35, Thm. 7.5.1], a strengthening we shall use in 2.17 (and only there).

2.14 Theorem.   *Let f be $\mathcal{Z}$-finite and K-finite, and let U be a neighborhood of 1. Then there exists $\alpha \in I_c^\infty(G)$ with support in U such that*

$$f = f * \alpha.$$

This is a fundamental theorem proved (in much greater generality) by Harish-Chandra [29] (see also [7]). In this section, we reduce its proof to that of a statement in representation theory. Assume first f to be of type n. Denote by V the smallest G-invariant closed subspace of $C^\infty(G)$ containing f. Then it can be shown that its elements are annihilated by $P(\mathcal{C})$ and that

$$V_\chi = \{\, v \in V \mid v \text{ is of type } \chi \,\}$$

is finite dimensional for every $\chi \in \hat{K}$. This is the result we need. A proof is given in Sections 2.17–2.21.

In general, f is a finite sum of elements f_i, where f_i is of type $n(i)$ for some $n(i) \in \mathbb{Z}$ (2.12). We see therefore that there exists a finite dimensional subspace L of $C^\infty(G)$ containing f, stable under the convolutions $*\alpha$ ($\alpha \in I_c^\infty(G)$) and containing therefore all elements $f * \alpha$. The convolutions $*\alpha$, restricted to L, form a vector subspace W of the space End L of endomorphisms of L that is necessarily finite dimensional. If $\{\alpha_n\}$ ($n = 1, 2, \ldots$) is a Dirac sequence, then $h * \alpha_n \to h$ for any $h \in V$ and hence $*\alpha_n$ tends to the identity. Thus the identity is in the closure of W and therefore belongs to W, since we deal with finite dimensional vector spaces. This proves the theorem.

2.15   The equality $f = f * \alpha$ means that $f(x)$ is some sort of average of the values of f over a neighborhood. This analogy can be made very precise in the case of harmonic functions.

Note first that the definition of $*$ makes sense in any Lie group, in particular on $\mathbb{R}^n$, viewed as a Lie group with respect to addition. There it is the classical convolution

$$(\alpha * \beta)(x) = \int \alpha(x + y)\beta(-y)\, dy = \int \alpha(y)\beta(x - y)\, dy.$$

Let B_1 be the unit ball centered at the origin and $\chi_\circ$ its characteristic function divided by the Euclidean volume $v(B_1)$ of B_1. Then

$$(f * \chi_\circ)(x) = \int_{\mathbb{R}^n} f(y)\chi_\circ(x - y)\, dy = v(B_1)^{-1} \int_{x+B_1} f(y)\, dy.$$

Hence $f = f * \chi_\circ$ means that the value of f at x is the average of its values on the unit ball centered at x. This is the case if f is harmonic.

In the remainder of this section we supply a proof of the two statements used in 2.14. As the reader will see from the proof and from the references to more general results, its natural framework is that of representation theory. We basically stick to $C^\infty(G)$ in order to minimize the prerequisites.

2.16 Lemma.   *Let $f \in C^\infty(G)$ be of right K-type χ ($\chi \in \hat{K}$), and let $D \in \mathcal{U}(\mathfrak{g})$. Then Df is of finite K-type on the right.*

Proof. It is enough to prove this when $D = X_1 \ldots X_s$ ($X_i \in \mathfrak{g}$, $1 \leqslant i \leqslant s$). We have (see 2.1(6))

$$(r_k Df)(x) = Df(xk) = \frac{d^s}{dt_1 \dots dt_s} f(xke^{t_1 X_1} \dots e^{t_s X_s}) \bigg|_{t_1 = t_2 = \dots = t_s = 0}.$$

Now (see 2.1(3))

$$f(xke^{t_1 X_1} \dots e^{t_s X_s}) = f(xe^{t_1{}^k X_1} e^{t_2{}^k X_2} \dots e^{t_s{}^k X_s} k) = \chi(k) f(xe^{t_1{}^k X_1} \dots e^{t_s{}^k X_s});$$

hence

$$(r_k Df)(x) = \chi(k)({}^k X_1{}^k X_2 \dots {}^k X_s) f(x).$$

If A, B, C is a basis of $\mathfrak{g}$, then ${}^k X_1 \dots {}^k X_s$ is a linear combination, with real coefficients (depending on k), of products of s factors each equal to one of A, B, C. There exist therefore finitely many elements $D_1, \dots, D_N$ of $\mathcal{U}(\mathfrak{g})$ such that $r_k Df$ is a linear combination, with constant coefficients, of the $D_i f$ $(i = 1, \dots, N)$ for each $k \in K$.

2.17 Theorem. *Let* $f \in C^\infty(G)$ *be* $\mathcal{Z}$-*finite and of* K-*type* n. *Let* $P(\mathcal{C})$ *be a monic polynomial in the Casimir operator* $\mathcal{C}$ *that annihilates* f. *Then*:

(a) *the closure* V *in* $C^\infty(G)$ *(in the* C^∞-*topology) of* $\mathcal{U}(\mathfrak{g}).f$ *is* G-*invariant*; *and*

(b) *for every* $\chi \in \hat{K}$, *the space*

$$V_\chi = \{ v \in V \mid r_k v = \chi(k) v, \; k \in K) \}$$

is finite dimensional.

Note that (a) implies the space V considered in 2.14 is the closure of $U(\mathfrak{g}).f$. Since $P(\mathcal{C})$ commutes with $\mathcal{U}(\mathfrak{g})$, every element of $\mathcal{U}(\mathfrak{g}).f$ is annihilated by $P(\mathcal{C})$, and hence so is every element of V. This and (b) are the two facts used in 2.14. It remains to prove (a) and (b).

Proof of (a). Let U be the smallest G-invariant closed subspace of $C^\infty(G)$ containing $\mathcal{U}(\mathfrak{g}).f$. It contains V and we must show that, in fact, $U = V$. By the Hahn–Banach theorem (see e.g. [46, III; 61, IV]) it suffices to show that, if b is any continuous linear form on U that vanishes on V and if $v \in V$, then b vanishes on $r_g v$ for any $g \in G$.

On G, consider the function $\varphi: g \mapsto b(r_g v)$. We claim first that φ is annihilated by $P(\mathcal{C})$ and is right K-finite. The first assertion is clear since $\mathcal{C}$ commutes with right or left translations. By 2.16, there exists $v_1, \dots, v_s \in \mathcal{U}(\mathfrak{g}).f$ such that $r_k v$ is a finite linear combination of the v_is. Then φ is a linear combination of the functions $g \mapsto b(r_g.v_i)$ and hence is also K-finite. By the remark in 2.13, φ is analytic. For any $D \in \mathcal{U}(\mathfrak{g})$, we have $D\varphi(1) = b(Dv) = 0$. Therefore φ is an analytic function all of whose derivatives at 1 vanish (see 2.1); hence $b(r_g v) = 0$ for all $g \in G$. This concludes the proof of (a).

2.18   To prove (b), the crux of the matter is the purely algebraic theorem 2.19 on $\mathcal{U}(\mathfrak{g}).f$. We first make some preparation for its proof.

Let $G_c = SL_2(\mathbb{C})$ be the group of complex 2×2 matrices of determinant 1. Its Lie algebra $\mathfrak{g}_c$ is the Lie algebra of 2×2 complex matrices of trace zero. It is elementary that there exists $g \in G_c$ such that $g.H.g^{-1} = i.W$. Let then $Y = g.E.g^{-1}$ and $Z = g.F.g^{-1}$. Since Ad g is an automorphism of $\mathfrak{g}_c$, we have, as in 2.1(1),

(1) $[iW, Y] = 2Y, \quad [iW, Z] = -2Z, \quad [Y, Z] = iW.$

Note that since $\mathcal{U}(\mathfrak{g})$ is an algebra over $\mathbb{C}$, it may be identified with $\mathcal{U}(\mathfrak{g}_c)$. It is spanned by the products

(2) $Y^a.Z^b.(iW)^c \quad (a, b, c \in \mathbb{N}).$

As in 2.5(1), we have

(3) $\mathcal{C} = -\frac{1}{2}W^2 + YZ + ZY = -\frac{1}{2}W^2 + iW + 2ZY = -\frac{1}{2}W^2 - iW + 2YZ.$

Now let M be a complex vector space that is a $\mathcal{U}(\mathfrak{g})$-module. An element $v \in M$ is said to be *of weight λ with respect to iW* if $iW.v = \lambda v$. The set of such elements is the *weight space M_λ*. We claim that if v is of weight λ then $Y^a.Z^b.v$ is of weight $\lambda + 2(a - b)$. By (1) we can write

(4) $iW.Y = Y.iW + 2Y, \quad iW.Z = Z.iW - 2Z;$

hence

(5) $iW.Yv = (\lambda + 2)Y.v, \quad iW.Zv = (\lambda - 2)Zv,$

so that our assertion follows immediately by induction on a and b.

A $\mathcal{U}(\mathfrak{g})$-module M is said to be *cyclic* if it is generated by some element over $\mathcal{U}(\mathfrak{g})$, that is, if $M = \mathcal{U}(\mathfrak{g}).v$ for some $v \in M$.

2.19 Theorem.   *Let $M = \mathcal{U}(\mathfrak{g}).v$ be a cyclic $\mathcal{U}(\mathfrak{g})$-module generated by one element v that is $\mathcal{Z}$-finite and an eigenvector of iW with weight λ ($\lambda \in \mathbb{C}$). Then M is a countable direct sum of weight spaces M_μ for iW, with $\mu \in \lambda + 2\mathbb{Z}$. If v is an eigenfunction of $\mathcal{C}$ then $\dim M_\mu \leqslant 1$ ($\mu \in \mathbb{C}$).*

(In the general case of a module over $\mathcal{U}(\mathfrak{h})$, where $\mathfrak{h}$ is a semisimple Lie algebra, this result follows from [28, Thm. 1].)

Proof. The line $\mathbb{C}.v$ is stable under all powers of iW, so M is spanned over $\mathbb{C}$ by the elements $Y^a.Z^b.v$ $(a, b \in \mathbb{N})$. Since the latter vector is of weight $2(a - b) + \lambda$ (see 2.18), this shows that M is a direct sum of spaces M_μ with $\mu \in \lambda + 2\mathbb{Z}$.

By assumption, v is annihilated by some nonconstant monic polynomial $P(C)$ in the Casimir operator C. Note first that since $P(C)$ is in the center of $\mathcal{U}(\mathfrak{g})$, it annihilates all elements of M. To prove the finite dimensionality of the M_μ, we first consider the case where v is an eigenvector of C, that is, where $P(C) = C - c$ for some $c \in \mathbb{C}$. In that case we want to prove that M_μ is at most one-dimensional. We first claim:

(1) *If* $x \in M_\mu$, *then* $Y^a.Z^a.x \in \mathbb{C}.x$ *for all* $a \in \mathbb{N}$.

The proof is by induction on a. By 2.18(3), YZ is a polynomial in iW and C:

$$YZ = Q(iW, C) = \tfrac{1}{2}C + \tfrac{i}{2}W + \tfrac{1}{4}W^2;$$

therefore $YZ.x = Q(\mu, c).x$. Similarly, x is an eigenvector of $Z.Y$. Assume our assertion proved up to a. We have

$$Y^{a+1}.Z^{a+1} = Y.X^a.Z^a.Z.$$

Now $Z.x \in M_{\mu-2}$ (see 2.18), so $Y^a.Z^a.Z.x$ is a multiple of $Z.x$ by the induction assumption. Thus $Y^{a+1}Z^{a+1}x$ is a multiple of YZx and hence of x. This proves (1). It follows that if $a \geqslant b$ then

$$Y^a.Z^b.v = Y^{a-b}.Y^b.Z^b.v \in \mathbb{C}Y^{a-b}.v \in M_{\lambda+2(a-b)}$$

and, if $a \leqslant b$, then

$$Y^a.Z^b = Y^a.Z^a.Z^{b-a}v \in \mathbb{C}Z^{b-a}.v \in M_{\lambda+2(a-b)}.$$

Thus M is spanned by the vectors $Y^m v$ and $Z^m v$ of respective weights $\lambda + 2m$ and $\lambda - 2m$ ($m \in \mathbb{N}$). Therefore the weights belong to $\lambda + 2\mathbb{Z}$ and have multiplicity 1. This establishes the theorem when $P(C) = C - c$. To prove the finite dimensionality of the M_μ in the general case, we argue by induction. Assume first that $P(C) = (C - c)^a$ ($a \geqslant 2$) and that our assertion is proved if $P(C) = (C - c)^{a-1}$. We consider the following short exact sequence of cyclic $\mathcal{U}(\mathfrak{g})$-modules

(2) $0 \rightarrow \mathcal{U}(\mathfrak{g}).(C - c).v \rightarrow M \rightarrow M/\mathcal{U}(\mathfrak{g}).(C - c).v.$

We have already seen that these modules are direct sums of eigenspaces with respect to iW, so (2) also yields exact sequences

(3) $0 \rightarrow (\mathcal{U}(\mathfrak{g}).(C - c)v)_\mu \rightarrow M_\mu \rightarrow (M/\mathcal{U}(\mathfrak{g}).(C - c)v)_\mu$ ($\mu \in \mathbb{C}$).

The image of v in the last term of (2) generates it over $\mathcal{U}(\mathfrak{g})$ and is annihilated by $(C - c)$. The left-hand module in (2) is generated by $(C - c).v$, which is annihilated by $(C - c)^{a-1}$. Therefore, the two extreme terms of (3) are finite dimensional and hence so is the middle one. Also, if $M_\mu \neq 0$ then μ is a weight of either the left-hand module or the right-hand one; hence $\mu \in \lambda + 2\mathbb{Z}$ by induction.

In the general case, we can write

(4)　$P(C) = (C - c_1)^{a_1} \ldots (C - c_s)^{a_s}$　$(a_i \in \mathbb{N}, i = 1 \ldots s; c_i \neq c_j \text{ if } i \neq j).$

Let $P_i(C) = P(C).(C - c_i)^{-a_i}$. The $P_i(C)$ have no common nonconstant factor, so there exist polynomials $Q_i(C)$ such that

(5)　　　　　　　　　　　$1 = \sum_{i=1}^{i=s} Q_i(C) P_i(C).$

Let $v_i = Q_i(C).P_i(C)v$. Then v is the sum of the v_is and hence M is the sum of the submodules $M_i = \mathcal{U}(\mathfrak{g}).v_i$. This reduces us to proving 2.17 for each M_i. Since $(C - c_i)^{a_i}.v_i = Q_i(C)P(C).v = 0$, we are back to the previous case.

2.20　We now take for M the module $\mathcal{U}(\mathfrak{g}).f$ of 2.17 and consider the weights of W rather than of iW. The weight of f being equal to im, we see that the weights of W in M belong to $i\mathbb{Z}$. In order to prove 2.17(b), it suffices to show, in view of 2.19, that

(1)　　　　　　　$V_m = M_{im}$　where　$V_m := V_{\chi_m}.$

To this effect, we introduce here some formalism of much more general scope (to which we shall return in 14.4). We let dk be the Haar measure on K of total mass 1 – that is, $d\varphi/2\pi$, in the parameterization of K introduced in 2.11, where K is identified with $\{e^{i\varphi}\}$. We have the familiar relation

(2)　　　　　　　$\int_0^{2\pi} e^{im\varphi}.e^{-in\varphi} \, d\varphi = 2\pi\delta_{m,n},$

which can be written as

(3)　　　　　　　$\int_K \chi_m(k)\chi_{-n}(k) \, dk = \delta_{m,n}.$

Given $h \in C^\infty(G)$, we can, for each $x \in G$, consider the function h_x on K defined by $k \mapsto h(x.k)$ and then the convolution $h_x * \chi_m$ on K. The resulting function $h_m : x \mapsto h_x * \chi_m$ will also be denoted $h * \chi_m$. It is indeed a convolution on G if we identify χ_m as the measure on X with support on K and equal to $\chi_m \, dk$ on K. This function is called the mth *Fourier component* of h (with respect to K). We therefore have

(4)　　　$h_m(x) = \int_K h(x.k)\chi_m(k^{-1}) \, dk = \int_K h(x.k).\overline{\chi_m(k)} \, dk.$

Assume that h is of type n. Then, using (3), we obtain

(5)　　　　　$h_m(x) = \int h(x)\chi_n(k)\chi_m(k^{-1}) \, dk = h(x)\delta_{m,n},$

hence

(6) $h_n = \delta_{m,n} h$ (*h* of type *m*).

This shows that $*\chi_n$ is a projector of $C^\infty(G)$ onto the space of elements of *K*-type *n* (on the right). By standard Fourier analysis, the Fourier series $\sum h_n$ converges absolutely to *h*, but we shall not need this fact.

2.21 *Proof of 2.17(b):* We need to establish 2.20(1). Since M_{im} is finite dimensional, it is closed in *V*. It suffices therefore to show that any $v \in V_m$ is a limit of elements in M_{im}.

The space *M* is the direct sum of M_{im} and of the direct sum $M(im)$ of the M_{in} ($n \neq m$), hence *V* is the direct sum of M_{im} and of the closure $V(im)$ of $M(im)$. It follows from 2.20(6) that $*\chi_m$ annihilates $M(im)$, and hence also $V(im)$, since it is obviously continuous.

Let now $v \in V_m$. There exist $a_j \in M_{im}$ and $b_j \in M(im)$ ($j = 1, \ldots$) such that $v = \lim_{j \to \infty}(a_j + b_j)$. By 2.20(6), then, we have

$$v = v * \chi_m = \lim_j(a_j * \chi_m + b_j * \chi_m) = \lim_j a_j$$

as was to be proved.

2.22 Corollary. *Let $p \geqslant 1$. Let f be $\mathcal{Z}$-finite, K-finite on the right or on the left, and in $L^p(G)$. Then f is bounded on G and belongs to $L^r(G)$ for all $r \geqslant p$.*

Proof. Assume *f* to be right *K*-finite. By 2.14, there exists $\alpha \in I_c^\infty(G)$ such that $f = f * \alpha$. Then, by Hölder's inequality

(1) $|f(x)| \leqslant \int_G |f(x.y).\alpha(y^{-1})|\,dy \leqslant \|l_x f\|_p.\|\alpha\|_q = \|f\|_p.\|\alpha\|_q,$

where $q^{-1} = 1 - p^{-1}$; therefore, $|f|$ is bounded. If *f* is left *K*-finite, then $f = \alpha * f$ ($\alpha \in C_c^\infty(G)$) and the proof is similar. That it then belongs to $L^r(G)$ for all $r \geqslant p$ follows from the following elementary lemma.

2.23 Lemma. *Let $p \geqslant 1$. Let $f \in C(G)$ be bounded and belong to $L^p(G)$. Then $f \in L^r(G)$ for all $r \geqslant p$.*

Proof. Write $G = D \cup E$, where

$$D = \{x \in G, |f(x)| \leqslant 1\}, \qquad E = \{x \in G, |f(x)| \geqslant 1\}.$$

The relation $f \in L^p(G)$ implies that $\int_E dx < \infty$. The function *f* being bounded, so is f^p and hence $\int_E |f(x)|^p\,dx < \infty$. On *D* we have $|f(x)|^r \leqslant |f(x)|^p$, whence

$$\int_D |f(x)|^r\,dx \leqslant \int_D |f(x)|^p\,dx \leqslant \int_G |f(x)|^p\,dx < \infty.$$

3 Action of G on $\bar{\mathsf{X}}$. Discrete subgroups of G. Reduction theory

In this section, we review some basic facts about the action of G on the closure $\bar{\mathsf{X}}$ of X and about fundamental domains of discrete subgroups. This material is classical. It is collected here for the sake of completeness and to fix notation.

3.1   We recall that

$$PG = P\,\mathrm{SL}_2(\mathbb{R}) = \mathrm{SL}_2(\mathbb{R})/\{\pm 1\}$$

is the group Aut X of conformal transformations of X. The subgroup $K/\{\pm 1\}$ is the isotropy subgroup of $i: gi = i \iff g \in K$. We have

$$\begin{pmatrix} 1 & x \\ 0 & 1 \end{pmatrix} \cdot \begin{pmatrix} y^{1/2} & 0 \\ 0 & y^{-1/2} \end{pmatrix} . k(i) = x + yi \quad (y > 0,\ k \in K).$$

Hence the map $g \mapsto g.i$ yields a diffeomorphism π:

$$NA \cong \mathsf{X} = G/K,$$

which maps

$$\begin{pmatrix} 1 & x \\ 0 & 1 \end{pmatrix} \cdot \begin{pmatrix} y^{1/2} & 0 \\ 0 & y^{-1/2} \end{pmatrix}$$

onto $x + yi$ $(x, y \in \mathbb{R},\ y > 0)$.

3.2   The action of G on X extends continuously to $\mathsf{X} \cup \mathbb{R} \cup \infty$. We let

$$\bar{\mathsf{X}} = \mathsf{X} \cup \mathbb{R} \cup \infty,$$

which is diffeomorphic to the closed unit disc. We write also $\partial \bar{\mathsf{X}}$ for $\mathbb{R} \cup \infty$. We have $g(\infty) = a/c$; therefore,

$$g(\infty) = \infty \iff c = 0.$$

If $x \in \mathbb{R}$, then $g(x) = \infty \iff cx + d = 0$. Moreover, $G_x = g^{-1}G_\infty g$ is conjugate to G_∞. Thus the parabolic subgroups of G (see 2.6) are the isotropy groups of the points of $\partial \bar{\mathsf{X}}$. The group K acts transitively on $\partial \bar{\mathsf{X}}$. More precisely, $K/\{\pm 1\}$ is simply transitive on $\partial \bar{\mathsf{X}}$.

If P fixes $p \in \partial \bar{\mathsf{X}}$, then the Cartan subgroups of P are the connected subgroups of P fixing p and another point of $\partial \bar{\mathsf{X}}$.

3.3   We review some elementary facts about X, considered as a Riemannian manifold. The group PG is also the group of *orientation preserving isometries* of X with respect to the Riemannian metric of Gaussian curvature -1 given by

(1)
$$ds^2 = (dx^2 + dy^2).y^{-2},$$

with associated invariant volume element

(2)
$$dv = (dx \wedge dy).y^{-2}.$$

The (complete) geodesics are the half-circles orthogonal to $\mathbb{R}$ and the vertical lines. Each geodesic has two accumulation points on $\partial\bar{X}$, to be called its *endpoints*. A geodesic is stable under the Cartan subgroup leaving its endpoints fixed, and is the only geodesic having this property. Let $p \in \partial\bar{X}$ and let $P = G_p$ be its isotropy group, $N = N_P$. The group P permutes the geodesics having p as one endpoint. Its orbits, called *horocycles*, are the orthogonal trajectories to those geodesics. If $p \neq \infty$, the horocycles are the circles in $\bar{X}$ tangent to $\mathbb{R}$ at p. The closed disc bounded by a horocycle will be called a *horodisc*. If $p = \infty$, the geodesics ending at p are the vertical lines; the corresponding horocycles are the horizontal lines $\mathbb{R} + i.t$ $(t > 0)$. The region $y > t$ will be called a *horostrip*. (Of course, it can be viewed as a horodisc limited by a circle of infinite radius. If we use the unit disc model of §4, all points of $\partial\bar{X}$ are treated in the same way and so this distinction would disappear. However, for most computations at a cusp, it is more convenient to consider ∞ in the upper half-plane model.) It is clear that the intersection of two horodiscs tangent at different points of $\partial\bar{X}$ is compact. Hence, so is the intersection of their inverse images with respect to $\pi: G \to X$.

It is elementary to check that

(3)
$$\mathcal{I}(gz) = \mathcal{I}z.|cz + d|^{-2} \quad \left(g = \begin{pmatrix} a & b \\ c & d \end{pmatrix} \right).$$

We have

(4)
$$|cz + d|^2 = (cx + d)^2 + c^2.y^2 \quad (z = x + iy);$$

therefore,

(5)
$$\mathcal{I}z.\mathcal{I}(gz) \leqslant c^{-2} \quad \text{if } c \neq 0.$$

In the sequel, we let $\mu(g, z)$ be the function on $G \times X$ defined by

(6)
$$\mu(g, z) = cz + d \quad \left(g = \begin{pmatrix} a & b \\ c & d \end{pmatrix} \right).$$

This function is an *automorphy factor*, that is, it satisfies the cocycle identity

(7)
$$\mu(g.g', z) = \mu(g, g'.z).\mu(g', z) \quad (g, g' \in G, z \in X).$$

We recall, finally, that the Laplace–Beltrami operator for the metric (1) is

(8)
$$\Delta = -y^2 \left(\frac{\partial^2}{\partial x^2} + \frac{\partial^2}{\partial y^2} \right);$$

therefore (see 2.5),

$$(9) \qquad\qquad C^* = -2.\Delta.$$

3.4　Let H be a locally compact group acting continuously on a locally compact space V. Then H is said to act *properly* if, given $C \subset V$ compact, the set $\{ h \in H \mid hC \cap C \neq \emptyset \}$ is compact and, in particular, finite if H is discrete, in which case the action of H on V is often said to be *properly discontinuous*.

Assume that H acts properly on V; then so does every closed subgroup. The space V/H of orbits of H in V, endowed with the quotient topology, is locally compact and, in particular, Hausdorff. (For all this see e.g. [11, §§I.10, III.4].)

If $V = H/L$, where L is a compact subgroup of H, then it is clear from the definitions that H, operating by left translations on V, acts properly. In particular G acts properly on X (but not on $X̄$, though, since the isotropy groups on $\partial X̄$ are not compact).

3.5　An element $g \in G$ ($g \neq \pm 1$) is said to be *elliptic* (resp. *hyperbolic*, *parabolic*) if it is conjugate to an element of K (resp. A, N). The following equivalences are elementary:

(a) g is elliptic $\Longleftrightarrow$ the eigenvalues of g are complex of modulus 1 and distinct $\Longleftrightarrow$ tr $g < 2 \Longleftrightarrow g$ has exactly one fixed point on X and none on $\partial X̄$;

(b) g is hyperbolic $\Longleftrightarrow$ the eigenvalues of g are real distinct $\Longleftrightarrow$ tr $g > 2 \Longleftrightarrow g$ has two fixed points on $\partial X̄$ and none on X;

(c) g is parabolic $\Longleftrightarrow$ the eigenvalues of g are equal to 1 $\Longleftrightarrow$ tr $g = 2 \Longleftrightarrow g$ has one fixed point on $\partial X̄$ and none on X.

The image of g in PG will also be called elliptic or hyperbolic or parabolic as g is. Note that every element in PG is of one of these types; G itself is the union of the sets of elliptic, hyperbolic, and parabolic elements with the conjugates of $-A.N$.

An element $g \in G$ is semisimple (i.e. diagonalizable over $\mathbb{C}$) if and only if it is elliptic, hyperbolic, or equal to ± 1. It is unipotent $\neq 1$ if and only if it is parabolic. The conjugacy class of an elliptic or hyperbolic element consists of all the elements in G with the same eigenvalues, which are distinct, and hence is closed in G. If $g \in CG$, its conjugacy class reduces to g and hence is also closed. The conjugacy class of a parabolic element is not closed and in fact has 1 as accumulation point. It suffices to see this for $g = \begin{pmatrix} 1 & x \\ 0 & 1 \end{pmatrix}$, $x \neq 0$. For $a \in A_0$ diagonal, we have

$$a.g.a^{-1} = \begin{pmatrix} 1 & a^{2\rho}.x \\ 0 & 1 \end{pmatrix}$$

(see 2.8 for the notation), which tends to the identity if $a^{2\rho} \to 0$.

3.6 Isotropy subgroups in Γ. Let $z \in \bar{\mathsf{X}}$. As usual, Γ_z denotes the isotropy group of z in Γ, that is, the subgroup of elements in Γ leaving z fixed. Let also Γ'_z be the isotropy group of z in the image Γ' of Γ in PG. We are interested in the structure of Γ_z or Γ'_z when $\Gamma'_z \neq \{1\}$. There are three cases.

(a) $z \in \mathsf{X}$. Then Γ_z is isomorphic to a discrete subgroup of K and hence is finite cyclic, generated by an elliptic element. The point z is said to be *elliptic* for Γ.

Let now $z \in \partial\bar{\mathsf{X}}$. Its isotropy group in G is a parabolic subgroup P. Let $N = N_P$.

(b) The point z is said to be *cuspidal* for Γ, or Γ-*cuspidal*, if $\Gamma_N = \Gamma \cap N \neq \{1\}$. Then Γ_N is infinite cyclic, generated by a parabolic element. We claim that Γ_N has index $\leqslant 2$ in $\Gamma_P := \Gamma \cap P$. If the index is 2, then either $\Gamma_P = CG.\Gamma_N$ or Γ_P is infinite cyclic, generated by an element in $-N$. In both cases, Γ'_z is equal to Γ'_N and is infinite cyclic.

Proof. By 2.8(7), an element $p \in P$ acts on N or its Lie algebra by the dilation by $h_P(p)^{2\rho}.I_d$. The group Γ_N is invariant under Γ_P and so any $p \in \Gamma_P$ induces an automorphism on N/Γ_N, which is the circle group, whence $h_P(p)^{2\rho} = 1$. As a consequence, the projection of Γ_P on $\pm A$ (modulo N) is equal to ± 1, and our assertion follows.

(c) Assume $\Gamma_N = \{1\}$. The projection of P' onto A is then injective on Γ'_z, hence the latter is commutative. The group P' is the union of N' and of its Cartan subgroups, and each Cartan subgroup is its own centralizer in P'. Therefore Γ'_z is contained in one of them and so is infinite cyclic, generated by a hyperbolic element. There exists then a Cartan subgroup A of P such that $\Gamma_P \subset A \cup -A$. Then z is said to be *hyperbolic* for Γ.

The group Γ is discrete in G and thus countable, so the sets of elliptic or parabolic or hyperbolic points for Γ are also countable. Moreover, the set of elliptic points for Γ is discrete on X, since Γ acts properly on X. Therefore, the set of points $x \in \mathsf{X}$ with $\Gamma_z \subset CG$ is open and dense.

3.7 Assume that ∞ is a cuspidal point for Γ. Then (see 3.5) there exists $h > 0$ such that $\Gamma_\infty \subset (CG).\Gamma_0$, where Γ_0 is generated by $\begin{pmatrix} 1 & h \\ 0 & 1 \end{pmatrix}$. Let $\gamma = \begin{pmatrix} a & b \\ c & d \end{pmatrix}$. We also write $c(\gamma)$ for c. An elementary matrix computation shows that $|c(\gamma)|$ is constant on the double cosets $\Gamma_\infty \sigma \Gamma_\infty$ ($\sigma \in \Gamma$).

Lemma. *Let $M > 0$. Then there exist only finitely many double cosets $\Gamma_\infty \sigma \Gamma_\infty$ in Γ such that $|c(\gamma)| \leqslant M$ for $\gamma \in \Gamma_\infty \sigma \Gamma_\infty$. There exists $r > 0$ such that $|c(\gamma)| \geqslant r$ for all $\gamma \in \Gamma - \Gamma_\infty$.*

Proof. Since Γ_∞ is the set of $\gamma \in \Gamma$ such that $c(\gamma) = 0$, the second assertion follows from the first one. Let $\gamma_0 = \begin{pmatrix} 1 & h \\ 0 & 1 \end{pmatrix}$. For $m, n \in \mathbb{N}$ let

(1) $$\gamma_0^m \cdot \gamma \cdot \gamma_0^n = \begin{pmatrix} a' & b' \\ c' & d' \end{pmatrix}.$$

Then

(2) $$c = c', \qquad d' = d + n.h.c.$$

There exists n such that

(3) $$1 \leqslant d + n.h.|c| \leqslant 1 + h.|c|;$$

therefore,

(4) $$\mathcal{I}\left(\frac{a'i + b'}{c'i + d'}\right) = (c'^2 + d'^2)^{-2} \in [(M^2 + (1 + hM)^2)^{-1}, 1].$$

On the other hand, we may find m such that

(5) $$\left|\mathcal{R}\left(\frac{a'i + b'}{c'i + d'}\right)\right| \leqslant h.$$

This implies that any double coset on which $|c| \leqslant M$ has a representative γ' such that $\gamma'(i)$ is contained in a given compact subset. Since Γ acts properly, the set of such γ' is finite, which proves the first assertion and the lemma.

Note that 3.2 and 3.3(5) imply

(6) $$\mathcal{I}z.\mathcal{I}(\gamma z) < r^{-2} \quad (z \in H, \; \gamma \in \Gamma - \Gamma_\infty).$$

3.8 Lemma.   *Let u be a cuspidal point for Γ. Then there is a horodisc U tangent to $\partial \bar{X}$ at u such that*

$$\gamma.U \cap U \neq \emptyset, \quad \gamma \in \Gamma \Rightarrow \gamma \in \Gamma_u.$$

We may assume $u = \infty$. Let r be as in 3.7 and let U be the horostrip $y > 1/r$. Then, by 3.7(6),

$$\mathcal{I}(\gamma.z) < 1/r \quad (z \in U, \; \gamma \in \Gamma - \Gamma_\infty);$$

hence

$$U \cap \gamma U = \emptyset \quad \text{if } \gamma \in \Gamma - \Gamma_\infty.$$

3.9 The horodisc topology.   We let X_Γ^* be the union of X and the cuspidal points for Γ. We introduce on X_Γ^* a topology, to be called the *horodisc topology*. First, by definition, X is open in X_Γ^*. Let now u be a cuspidal point. As fundamental system of neighborhoods of u we take the union of u with horodiscs tangent to $\partial \bar{X}$ at u. On $\Gamma \backslash X_\Gamma^*$ we put the quotient topology. It is clear that the topology on X_Γ^* is Hausdorff, and that Γ acts by homeomorphisms on X_Γ^*. More generally, note first that if Γ' is commensurable with Γ (i.e. $\Gamma \cap \Gamma'$ has finite index in Γ and Γ'), then Γ and Γ' have the same cuspidal points. It suffices to see this when Γ' has finite index in Γ, in which case it follows from the definition (3.6). Let

$$\text{Com}(\Gamma) = \{\, g \in G \mid g\Gamma g^{-1} \text{ is commensurable with } \Gamma \,\}.$$

This is a subgroup of G called the *commensurator* or *commensurability group* of Γ which, by what has just been said, operates on X_Γ^*, obviously by homeomorphisms.

3.10 Proposition.

(i) *Let u be a cuspidal point for Γ and C a compact subset of X. Then there exists a horodisc U tangent at u such that $U \cap \Gamma(C) = \emptyset$.*

(ii) *The quotient space $\Gamma \backslash X_\Gamma^*$ is locally compact and, in particular, Hausdorff.*

Proof. (i) We may again assume that $u = \infty$. Let m, M be positive numbers such that

$$m \leqslant \mathcal{I}(z) \leqslant M \quad \text{for } z \in C.$$

If $\gamma \in \Gamma_\infty$ then $\mathcal{I}(\gamma z) = \mathcal{I}(z) \leqslant M$. If $\gamma \notin \Gamma_\infty$ then, with r as in 3.7, by 3.7(6) we have $\mathcal{I}(\gamma z) \leqslant m^{-1} . r^{-2}$. Therefore

$$\Gamma(C) \subset \{\, z \in X \mid \mathcal{I}z \leqslant \max(M, (mr^2)^{-1}) \,\},$$

and the horostrip $y > \max(M, (mr^2)^{-1})$ satisfies our conditions.

(ii) Let $\pi \colon X_\Gamma^* \to \Gamma \backslash X_\Gamma^*$ be the canonical projection, and let $u \in X_\Gamma^*$. We show first that $\pi(u)$ has a compact neighborhood. If $u \in X$ then this is clear since, by construction, $\Gamma \backslash X$ is open in $\Gamma \backslash X_\Gamma^*$. Let $u \in \bar{X}$ be cuspidal. Again we may assume that $u = \infty$. For $t > 0$ big enough, the equivalence relations defined by Γ_∞ and Γ on the horostrip H_t are the same (3.8), hence π maps $H_t \cup \{\infty\}$ continuously onto a neighborhood V of $\pi(U)$. The group Γ_∞ contains a translation by some $h > 0$. Therefore V is the image of $\{\infty\} \cup \{z \in H_t, |\mathcal{R}z| < h\}$, which is compact in the horodisc topology; hence V is compact. There remains to show that $\Gamma \backslash X_\Gamma^*$ is Hausdorff.

Let u and v be elements of X_Γ^* that are not in the same Γ-orbit. We must prove that u, v have neighborhoods U, V in the horodisc topology such that $U \cap \Gamma(V) = \emptyset$. Any point in X has a fundamental system of relatively compact neighborhoods. If $u, v \in X$ then our assertion follows because Γ acts properly on X. If $u \in X$ and $v \in \partial \bar{X}$, this follows from (i). There remains the case where

u, v are cuspidal for Γ. Again, we may assume that $u = \infty$. Then the union of u (resp. v) with the horostrips (resp. the horodiscs tangent to the real axis at v) form a fundamental system of neighborhoods of u (resp. v).

Let $h > 0$ be such that Γ_u contains the translation $z \mapsto z + h$. Fix $t > 0$ and let

$$(1) \qquad C = \{z = x + iy, z \in X, |x| \leqslant h, \text{ and } y = t\};$$

C is compact. By (i) there exists a horodisc V tangent to v such that

$$(2) \qquad V \cap \Gamma(C) = \emptyset.$$

Let U'_t be the horostrip $y > t$ and $U_t = U'_t \cup \{\infty\}$. We claim that the images of U and V in $\Gamma \backslash X^*_\Gamma$ are disjoint or, equivalently, that

$$(3) \qquad \Gamma(V) \cap U_t = \emptyset.$$

Let $\gamma \in \Gamma$ be such that $\gamma.V \cap U_t \neq \emptyset$. By assumption, $\gamma.v \neq \infty$, so $\gamma(V)$ is also a horodisc tangent to the real axis. Therefore $\gamma(V)$ intersects the horizontal line $y = t$. Since every point on $y = t$ is the image of an element in C by some $\gamma \in \Gamma_\infty$, it follows that there exists γ' such that $\gamma'(V) \cap C \neq \emptyset$, but this contradicts (2). This proves (3) and the proposition.

3.11 Some geometric definitions.   Our last goal in this section is to construct and establish some properties of a fundamental domain for Γ when $\Gamma \backslash G$ has finite volume or, equivalently, when $\Gamma \backslash X$ has finite hyperbolic area. We let $d(,)$ be the distance on X defined by the hyperbolic metric.

(a) If x, y, x', y' are four points such that $d(x, y) = d(x', y')$, then there exists $g \in G$ such that $gx = x'$ and $gy = y'$. Since G is transitive, it suffices to consider the case where $x = x' = i$. Then use the fact that the isotropy group K of i is transitive on the set of half geodesics passing through i, and therefore on the set of points having a given distance to i.

(b) Let $u, v \in X$. Then the set of points z such that $d(u, z) = d(v, z)$ is a geodesic that cuts the geodesic segment with endpoints u, v orthogonally at its middle point.

The geodesic segment with endpoints u, v is contained on a half-circle orthogonal to the real axis; by (a) we may assume that its middle point m is the highest point on that circle. Then it is easily checked that the locus of $d(z, u) = d(z, v)$ is the vertical line passing through m.

(c) Let C be a geodesic on X. Its complement consists of two open half-planes E_1, E_2 with boundary C. The subsets $E_i \cup C$ are closed half-planes. Following custom, and for lack of a better terminology, I shall call a *polygon* in X the intersection of such closed half-planes defined by a *locally finite* family of geodesics (i.e. a compact subset of X meets at most finitely many of them). If not empty, its interior is the intersection of the corresponding open half-planes. The polygon and its interior are *convex* (contain the unique geodesic segment joining any two of their points) since this is obviously the case for an open or closed half-plane.

3.12 Poincaré fundamental sets.   Recall that, if a group H acts on a set V, a fundamental domain Ω is a subset of V containing exactly one representative of each orbit. The projection $V \to H \backslash V$ on the space of orbits is then a bijection of Ω onto $H \backslash V$. Such an Ω always exists, at any rate if one is willing to use the axiom of choice. In our case, it is more useful to relax the condition on the fundamental domain in order to construct sets of representatives of orbits with good geometric properties. The conditions of interest are that the intersection of Ω with any H-orbit is finite, nonempty, or, more strongly, that

(0) $H.\Omega = V$ and $\{ h \in H \mid h\Omega \cap \Omega \neq \emptyset \}$ is finite.

We shall call (0) the *Siegel property*.

We recall Poincaré's construction of a fundamental set satisfying our first condition. Choose a point $z_\circ \in \mathsf{X}$ whose isotropy group is contained in CG – that is, whose isotropy group in the image Γ' of Γ in PG (see 3.1) is reduced to $\{1\}$. As pointed out in 3.6, these points form a dense open set. Given $\gamma \in \Gamma$ ($\gamma \neq \pm 1$), we let

(1) $E(\gamma, z_\circ) = \{ z \in \mathsf{X} \mid d(z, z_\circ) \leqslant d(z, \gamma.z_\circ) \}.$

This is a closed half-plane. Its boundary is the geodesic

(2) $F(\gamma, z_\circ) = \{ z \in \mathsf{X} \mid d(z, z_\circ) = d(z, \gamma.z_\circ) \}$

and its interior $E(\gamma, z_\circ)^\circ$ is defined by replacing $\leqslant$ by $<$ in (1). (See 3.11(c).) Let

(3) $$\Omega = \bigcap_{\gamma \in \Gamma - \{\pm 1\}} E(\gamma.z_\circ);$$

Ω is a non-empty polygon. Its interior is

(4) $$\Omega^\circ = \bigcap_{\gamma \in \Gamma - \{\pm 1\}} E(\gamma.z_\circ)^\circ.$$

Both Ω and Ω° are non-empty and convex, since they are intersections of convex sets.

The orbit $\Gamma z_\circ$ is discrete in X. Therefore, given a compact subset C of X, the intersection $C \cap \Omega$ is contained in all but finitely many $E(\gamma, z_\circ)$. The elements of $C \cap \Omega$ are defined by finitely many inequalities, and C meets only finitely many of the geodesics $F(\gamma, z_\circ)$.

Because the distance function is invariant under Γ, we see that

(5)
$$z \in \Omega \iff d(z, z_0) \leqslant d(z, \gamma z_0) \quad (\gamma \in \Gamma)$$
$$z \in \Omega^\circ \iff d(z, z_0) < d(z, \gamma z_0) \quad (\gamma \in \Gamma - \{\pm 1\}).$$

As a consequence, for any $u \in X$, the set $\Gamma u \cap \Omega$ is finite and non-empty, consists of points that are equidistant from z_0, and is reduced to one point if $\Gamma u \cap \Omega^\circ \neq \emptyset$.

In the sequel, we let $\bar{\Omega}$ be the closure of Ω in $X̄$; it is the disjoint union of Ω and of $\partial X̄ \cap \bar{\Omega}$. The points in the latter set will be called *points at infinity*. Since Ω is convex, it contains the unique geodesic arc with endpoints $x, y \in X$. By continuity, this remains true if one or both of x, y are at infinity.

3.13 Proposition. *The following conditions are equivalent*:

(i) $\Gamma \backslash G$ *is compact*;
(ii) $\Gamma \backslash X$ *is compact*;
(iii) Ω *is compact*;
(iv) $\bar{\Omega} \cap \partial X̄ = \emptyset$.

If these conditions are fulfilled, then $\partial \Omega$ consists of finitely many geodesic segments and Γ of semisimple elements.

Because K is compact, the equivalence of (i) and (ii) is obvious. The projection $\pi: X \to \Gamma \backslash X$ maps Ω onto $\Gamma \backslash X$. Therefore (iii) $\Rightarrow$ (ii). Assume (ii). Then there is a compact subset $C \subset X$ such that $\pi(C) = \Gamma \backslash X$. We may assume $z_0 \in C$. Let

$$c = \sup_{z \in C} d(z, z_0).$$

Given $u \in X$, Γu contains a point with distance $\leqslant c$ from z_0. Hence Ω is contained in a (hyperbolic) disc and so is compact; thus (ii) $\Rightarrow$ (iii). Obviously (iii) $\Rightarrow$ (iv). Assume that Ω is not compact. Then Ω contains a sequence of elements z_n $(n = 1, 2 \ldots)$ such that $d(z_n, z_0) \to \infty$. Passing to a subsequence if necessary, we may assume that it converges to some point $u \in X̄$. Any point in X has a neighborhood whose points have bounded distance from z_0. Therefore $u \in \partial X̄$. Thus (iv) $\Rightarrow$ (iii).

We have already pointed out that the geodesic segments on $\partial \bar{\Omega}$ form a locally finite family; they are therefore finite in number if Ω is compact. To prove the second part of the last assertion it suffices, by 3.5, to show that if $\Gamma \backslash G$ is compact then the conjugacy class in G of any element of Γ is closed in G. This follows from the following general simple principle.

Lemma. *Let H be a locally compact group, countable at infinity, and let L be a discrete cocompact subgroup. Then the conjugacy class $C_H(x)$ of any $x \in L$ is closed in H.*

Let $h_n \in H$ be such that $h_n.x.h_n^{-1}$ has a limit y. We need to show that y is conjugate to x. By assumption, $H = C.L$, where C is compact. We may write $h_n = c_n.l_n$, with $l_n \in L$ and $c_n \in C$. Passing to a subsequence, we may assume that $c_n \to c \in C$. Then $l_n.x.l_n^{-1} \to c^{-1}.y.c$. However, since L is discrete, the sequence $l_n.x.l_n^{-1}$ is stationary and so $l_n.x.l_n^{-1} = c^{-1}.y.c$ for a suitable n.

3.14 Theorem (Siegel). *Let Ω be a Poincaré fundamental set for Γ. Assume it has finite area. Then $\partial\Omega$ is the union of finitely many geodesic segments, $\partial\bar{\Omega} \cap \bar{X}$ is finite, and $\Gamma(\partial\bar{\Omega} \cap \bar{X})$ is the set of cuspidal points for Γ. The quotient $\Gamma\backslash X_\Gamma^*$ is compact, and Ω has the Siegel property.*

We first recall a basic fact in plane hyperbolic geometry. Let a, b, c be three points in X not contained on one geodesic; let D be the geodesic triangle with vertices a, b, c (i.e., the sides of D are the geodesic segments joining these three points); and let α, β, γ be the angles of D at a, b, c. Then the *hyperbolic area* of D is

$$(1) \qquad \mathrm{Ar}(D) = \pi - \alpha - \beta - \gamma.$$

The boundary $\partial\bar{\Omega}$ of Ω consists of geodesic arcs, possibly with endpoints at infinity (i.e., on $\partial\bar{X}$). Any vertex of $\partial\bar{\Omega}$ is the intersection of two geodesic arcs. If none is at infinity, then we are in the compact case and $\partial\Omega$ consists of finitely many arcs (3.13). So we assume that $\bar{\Omega}$ has at least one vertex at infinity; such vertices are the endpoints of connected components of $\partial\Omega$. To prove our first assertion, we assume that the number of geodesic segments in $\partial\Omega$ is infinite and derive a contradiction.

For each vertex a of $\partial\bar{\Omega}$, we let ω_a be the angle in the fundamental set of the two geodesics meeting at a. We have

$$(2) \qquad 0 \leqslant \omega_a < \pi,$$

where the value 0 is attained if and only if a is at infinity. We want to prove

$$(3) \qquad 2\pi + \mathrm{Ar}(\Omega) \geqslant \sum_a (\pi - \omega_a).$$

Consider one connected component S of $\partial\Omega$. Fix a vertex a_0 of S, and number the vertices consecutively $a_0, a_1, \ldots, a_n, \ldots$ in one direction and $a_0, a_{-1}, a_{-2}, \ldots,$ $a_{-m}, \ldots$ in the other. Let D_j be the geodesic triangle with vertices $z_\circ, a_j, a_{j+1}$ and $\alpha_j, \beta_j, \gamma_j$ its angles at $z_\circ, a_j, a_{j+1}$. Thus, if $a_j \in \partial\Omega$ then

$$(4) \qquad \omega_j = \beta_j + \gamma_{j+1}$$

is the angle ω_{a_j} in the notation of (2). Let $m \leqslant 0$ and $n \geqslant 0$ be such that a_m and a_n belong to the chain and are at a finite distance. The points a_{m-1} and a_{n+1} may or may not be at a finite distance. For $j \in [m-1, n]$ we have, by (1),

(5) $$\alpha_j + \text{Ar}(D_j) = \pi - \beta_j - \gamma_j.$$

Adding these equalities and taking (4) into account yields

(6) $$\sum_{m-1}^{n} \alpha_j + \sum_{m-1}^{n} \text{Ar}(D_j) = \pi - \beta_{m-1} - \gamma_n + \sum_{m}^{n} (\pi - \omega_j).$$

Assume first that the chain is finite and that a_{m-1} and a_{n+1} are at infinity. Then $\beta_{m-1} = \gamma_n = 0$ and we have

(7) $$\sum_{m-1}^{n} (\alpha_j + \text{Ar}(D_j)) = \pi + \sum_{m}^{n} (\pi - \omega_j).$$

Assume now that the chain is infinite – say, that n has no maximum – and let $n \to \infty$. The left-hand side of (6) is bounded by $2\pi + \text{Ar}(\Omega)$; hence the series on the right-hand side converges and so γ_j also has a limit, say γ_∞. We claim that $\gamma_\infty \leqslant \pi/2$. The sequence of distances $d(z_0, a_j)$ ($j = 0, 1, \ldots$) diverges, so there exist arbitrarily large j such that

$$d(z_0, a_{j+1}) > d(z_0, a_j).$$

This implies that $\beta_j > \gamma_j$ in the triangle D_j and, since $\beta_j + \gamma_j \leqslant \pi$, yields $\gamma_j \leqslant \pi/2$, whence our assertion. If a_{m-1} is at infinity, we have

(8) $$\sum_{m-1}^{n} (\alpha_j + \text{Ar}(D_j)) = \pi - \gamma_\infty + \sum_{m}^{\infty} (\pi - \omega_j) \geqslant \frac{\pi}{2} + \sum_{m}^{\infty} (\pi - \omega_j).$$

If the sequence of the a_j is also infinite in the other direction, then we see as before that β_j has a limit $\beta_{-\infty} \leqslant \pi/2$ as $j \to -\infty$, whence

(9) $$\sum_{-\infty}^{\infty} (\alpha_j + \text{Ar}(D_j)) \geqslant \sum_{-\infty}^{\infty} (\pi - \omega_j).$$

All the triangles with vertex z_0 considered here intersect at most on one side and cover Ω. Therefore, if we sum (7), (8), or (9), as the case may be, over all connected components of $\partial\Omega$, then the sum of the left-hand sides is the left-hand side of (3), and (3) follows.

The series on the right-hand side of (3) converges. There are only finitely many vertices such that, say,

(10) $$\omega_a < 9\pi/10.$$

Let C be the set of vertices that are Γ-equivalent to some element of the set defined by (10); C is finite (3.12). Because the number of vertices is assumed to be infinite, we can find a maximal set $b_1, \ldots, b_s$ of Γ-equivalent vertices outside C – that is, satisfying

(11) $9\pi/10 < \omega_{b_i} < \pi \quad (i = 1, \ldots, s).$

Let $\Omega = \Omega_1, \Omega_2, \ldots, \Omega_q$ be the set of Γ-transforms of Ω having b_1 as a vertex. Their angles at a are $< \pi$, hence $q \geqslant 3$. Let also c be the order of the isotropy group Γ'_{b_1} of b_1 in Γ' (i.e., modulo ± 1). Any $\gamma \in \Gamma'_b$ permutes the Ω_i. Given $j \in [1, q]$, there exists $\gamma \in \Gamma$ such that $\gamma(\Omega_j) = \Omega_1$. Then $\gamma(b_1)$ is one of the points b_i and the angle of Ω_i at a is ω_{b_i}. From this it follows that

(12) $\omega_{b_i} + \cdots + \omega_{b_q} = 2\pi/c,$

which is obviously incompatible with (11).

This contradiction shows that the number of geodesic arcs on $\partial\Omega$, and hence the number of vertices at finite distance, is finite. The number r of connected components of $\partial\Omega$ is equal to the cardinality of $\bar{\Omega} \cap \partial\bar{X}$. We are now only in the case (7) so that we get, by summation on the connected components of $\partial\Omega$,

(13) $2\pi + \mathrm{Ar}(\Delta) = r + \sum_a (\pi - \omega_a),$

which shows that r is finite.

We have already remarked that any Γ-orbit meets Ω° in at most one point. Therefore, if $\gamma\Omega \neq \Omega$ and $\gamma\Omega \cap \Omega \neq \emptyset$, this intersection is in $\partial\Omega$ and it is geometrically clear that it is either reduced to a vertex or contains a full geodesic arc of $\partial\Omega$. But such an arc can be only on two fundamental sets, and a vertex (in X) can be the vertex of at most finitely many $\gamma(\Omega)$. As a consequence,

$$\{\gamma \in \Gamma \mid \gamma\Omega \cap \Omega \neq \emptyset\}$$

is finite, that is, Ω has the Siegel property.

Let a be a vertex of $\partial\bar{\Omega}$ at infinity. We want to show that it is cuspidal. First let us prove that $\Gamma'_a \neq \{1\}$. The angle of Ω at a point at infinity is zero, so there are infinitely many transforms of Ω having a as a vertex. Given one of them, say Ω', there exists $\gamma \in \Gamma$ such that $\gamma(\Omega') = \Omega$. Then $\gamma(a)$ belongs to $\partial\bar{X} \cap \bar{\Omega}$. We can therefore find distinct fundamental sets Ω_1, Ω_2 with vertex a and $\sigma, \gamma \in \Gamma$ such that

$$\sigma(\Omega_1) = \gamma(\Omega_2) = \Omega \quad \text{and} \quad \sigma(a) = \gamma(a).$$

Then $\sigma^{-1}.\gamma \in \Gamma'_a - \{1\}$. We now claim that any $\gamma \in \Gamma'_a - \{1\}$ is parabolic. If not, then it is hyperbolic (3.5) and generates an infinite cyclic subgroup that is contained in some Cartan subgroup A fixing a and some other point b on $\partial\bar{X}$, hence also leaving invariant the geodesic L with endpoints a, b. Replacing γ by its inverse if necessary, we may arrange that the transforms of any z in L under the transformations γ^m ($m \geqslant 1$) accumulate to a. Thus the arc of L between a and any point c contains infinitely many Γ-equivalent points. Therefore (see 3.12), such an arc cannot be in any transform of Ω with vertex a – a contradiction. Consequently, $\Gamma'_a - \{1\}$ is infinite and consists of parabolic elements, and a is cuspidal for Γ.

Let $\pi: X^*_\Gamma \to \Gamma \backslash X^*_\Gamma$ be the canonical projection. In order to complete the proof of the theorem, it suffices to show that $\pi(\bar{\Omega})$ is compact and covers $\Gamma \backslash X^*_\Gamma$. Let a be a cuspidal point in $\bar{\Omega}$. The proof of 3.10(ii) shows that we can choose a sufficiently small horodisc D tangent to a such that the union of the intersections of D with finitely many transforms of $\bar{\Omega}$ with vertex a is mapped onto a compact neighborhood of $\pi(a)$. Because the intersection of D and such Ω' is equivalent modulo Γ to a subset of Ω, its projection belongs to $\pi(\Omega)$. Thus $\pi(\bar{\Omega})$ contains a compact neighborhood of the image of each point at infinity of $\bar{\Omega}$. For each such point a, let C_i be the intersection of Ω with a horodisc tangent at a. Then the complement of the union of the C_i in Ω is relatively compact in X and so its image under π is relatively compact (recall that $\pi(X)$ is open in $\Gamma \backslash X^*_\Gamma$). Thus $\pi(\bar{\Omega})$ is compact and contains $\pi(X)$. By definition of the horodisc topology (3.9), any non-empty open set on X^*_Γ intersects X in a non-empty open set. The complement of $\pi(\bar{\Omega})$ is open, does not meet $\pi(X)$, and hence is empty – whence our assertion.

Remark. The preceding proof is essentially Siegel's. He pushes the argument further to show that the minimum of the area of $\Gamma \backslash X$ is $\pi/21$. The minimum is achieved by a group with fundamental set a geodesic triangle having angles $2\pi/7, \pi/3, \pi/3$. The latter is compact. The minimum in the noncompact case is $\pi/3$, achieved by the modular group $SL_2(\mathbb{Z})$, which has one cusp. For the minimum in the case of more cusps, see 3.20.5.

3.15 Siegel sets. We now want to describe fundamental sets for Γ in a slightly different way that will be more convenient for many computations. It is based on a notion of a Siegel set, which has in fact already occurred implicitly.

(a) Let (P, A) be a (normal) p-pair (2.6) and N the unipotent radical of P. A *Siegel set* (with respect to (P, A)) in G is a subset of the form

$$(1) \qquad \mathfrak{S} = \omega. A_t. K, \quad \omega \subset N \text{ compact}, \quad A_t = \{a \in A \mid a^\alpha > t\}.$$

If $t' > t$ then the complement of $\omega. A_{t'}. K$ in $\mathfrak{S}$ is relatively compact in G.

The orbit $\mathfrak{S}' = \mathfrak{S}.i$ will be called a *Siegel set in* X. In both cases, subscripts P, t, ω may be added, depending on the amount of information deemed necessary in a given context. For brevity, we shall also say that $\mathfrak{S}'$ is a Siegel set at u or with respect to u, where $u \in \partial X$ is the fixed point of P, and we call $\mathfrak{S}'$ *cuspidal* for Γ if u is a cuspidal point for Γ. If $k \in K$, then $(^kP, ^kA)$ is a normal p-pair, $k.\mathfrak{S}$ is a Siegel set with respect to $(^kP, ^kA)$, and $k.\mathfrak{S}'$ is a Siegel set at $k(u)$. So, again, to investigate Siegel sets we may assume that $u = \infty$ and that (P, A) is the standard p-pair (see 2.6). Assume this, fix $h > 0$, and let

(2) $$\omega = \left\{ g = \begin{pmatrix} 1 & n \\ 0 & 1 \end{pmatrix} \middle| |n| \leqslant h \right\}.$$

Then

(3) $$\mathfrak{S}'_{t,\omega} = \{ z = x + iy \in X \mid |x| \leqslant h, \ y > t^2 \}.$$

If $\mathfrak{S}'$ is associated to u and ω is an interval, then $\mathfrak{S}'$ is the part of a horodisc tangent to $\partial\bar{X}$ at u contained between two geodesics with endpoint u. Accordingly it will also be called a *cusp at* u. Moreover, since we started from a normal p-pair, $\mathfrak{S}'$ contains an arc ending at u of the geodesic passing through i with endpoint u. Any translate $g.\mathfrak{S}'$ can be described geometrically as $\mathfrak{S}'$ and can be viewed as a cusp at $g(u)$. However, since K has been fixed once and for all (a convention usually made in harmonic analysis on reductive groups), it is more convenient to restrict ourselves to normal p-pairs. Thus a translate $g.\mathfrak{S}$ is in general *not* associated to a normal p-pair.

A Siegel set has *finite invariant measure*. It suffices to see this when $\mathfrak{S}'$ is associated to ∞. Then, in the notation of (2) and (3),

$$\mathrm{Ar}(\mathfrak{S}'_{P,t,\omega}) = \int_{\mathfrak{S}'} (dx \wedge dy).y^{-2} = \int_{\omega} dx \int_{y > t^2} dy.y^{-2} = t^{-2}. \int_{\omega} dx < \infty.$$

Similarly, a Siegel set in G has finite Haar measure.

(b) Assume that u is cuspidal for Γ and that ω contains an interval including a fundamental domain for Γ_u on N. Then

$$\Gamma_u.\mathfrak{S}' = H_{u,t}$$

is a horodisc tangent to $\partial\bar{X}$ at u. If $u = \infty$ then $\Gamma_u.\mathfrak{S}'$ is the horostrip

$$H_t = H_{\infty,t} = \{ z \in X \mid \mathcal{I}z > t^2 \}.$$

If t is big enough, the equivalence relations defined on $H_{u,t}$ by Γ and Γ_u are the same (3.8). Therefore, the projection $\pi: X_\Gamma^* \to \Gamma \backslash X_\Gamma^*$ is a finite-to-one mapping of $\{u\} \cup \mathfrak{S}'$ onto a relatively compact neighborhood of $\pi(u)$, which is equal to $\Gamma_u \backslash H_{u,t}$. The sets $\{u\} \cup \mathfrak{S}'$ form a fundamental set of neighborhoods of u in $\Gamma \backslash X_\Gamma^*$.

(c) If $\omega \subset N$ is relatively compact then so is $\bigcup_{a \in A_t} a^{-1}.\omega.a$, as follows from the relation $a^{-1}.n.a = a^{-2\rho}.n$ ($a \in A$, $n \in N$) (see 2.8(6)) and the boundedness of $a^{-2\rho}$ on A_t.

3.16 Lemma. *Let u and v be cuspidal for Γ. Let $\mathfrak{S}' = \mathfrak{S}'_{t,\omega}$ be a Siegel set at u and $\mathfrak{T}' = \mathfrak{T}'_{t',\omega'}$ a Siegel set at v.*

(i) *Let* $C \subset X$ *be relatively compact. Then* $\{\gamma \in \Gamma \mid \gamma(C) \cap \mathfrak{S}'_{t,\omega} \neq \emptyset\}$ *is finite.*

(ii) $\{\gamma \in \Gamma \mid \gamma(\mathfrak{S}') \cap \mathfrak{T}' \neq \emptyset\}$ *is finite. If* $v \notin \Gamma.u$, *then the intersection* $\gamma\mathfrak{S}' \cap \mathfrak{T}'$ *is* (a) *relatively compact for any* $\gamma \in \Gamma$ *and* (b) *empty if* $\gamma \notin \Gamma_u$ *and* t *is big enough.*

Proof. (i) We may assume that $u = \infty$. As in 3.10, let m and M be lower and upper bounds of $\mathcal{I}z$ for $z \in C$. Then (see 3.10(1)) there exists $c > 0$ such that

$$\Gamma(C) \subset \{z \in X \mid \mathcal{I}z \leqslant c\}.$$

Hence $\Gamma(C) \cap \mathfrak{S}'$ is contained in the set E of points in X with real part $\leq h$ in absolute value and imaginary part $> t^2$ and $\leq c$. The set E is relatively compact, so (i) follows from the fact that Γ acts properly.

(ii) We may again assume that $u = \infty$. First consider the case where $v \in \Gamma.u$. After translation by one suitable element in Γ, we may assume that $v = u = \infty$. For any $q > t$, $\mathfrak{S}'$ is the union of $\mathfrak{S}'_{q,\omega}$ and a relatively compact set C. By 3.8, we can choose q such that $\gamma(\mathfrak{S}'_{q,\omega})$ does not meet the horostrip H_q for any $\gamma \in \Gamma - \Gamma_\infty$. Then $\gamma(\mathfrak{S}'_{q,\omega}) \cap \mathfrak{T}'_{q,\omega'} = \emptyset$ for $\gamma \in \Gamma - \Gamma_\infty$. Thus, if $\gamma\mathfrak{S}' \cap \mathfrak{T}' \neq \emptyset$ then it follows that $\gamma(C) \cap D \neq \emptyset$, where $D = \mathfrak{T}'_{t,\omega'} - \mathfrak{T}'_{q,\omega'}$ is also relatively compact. The set of $\gamma \in \Gamma - \Gamma_\infty$ such that $\gamma(\mathfrak{S}') \cap \mathfrak{T}' \neq \emptyset$ is therefore finite. For $\gamma \in \Gamma_\infty$, the finiteness follows because $\{\gamma \in \Gamma_N, \gamma(\omega) \cap \omega' \neq \emptyset\}$ is finite.

Now assume that $v \notin \Gamma.u$. Because the space $\Gamma \backslash X_\Gamma^*$ is Hausdorff (3.13), we can find q such that $\Gamma(\mathfrak{S}'_{q,\omega}) \cap \mathfrak{T}'_{q,\omega'} = \emptyset$. Then, with the previous notation,

$$\{\gamma \in \Gamma \mid \gamma(\mathfrak{S}') \cap \mathfrak{T}' \neq \emptyset\} = \{\gamma \in \Gamma \mid \gamma(C) \cap D \neq \emptyset\},$$

which is finite, and also empty if q is big enough.

Let $\gamma \in \Gamma$ and assume that $\gamma(\mathfrak{S}') \cap \mathfrak{T}'$ is not compact. Then there exist sequences $t_n \to \infty$ and $x_n \in \mathfrak{S}'_{t_n}$ such that $\gamma x_n \in \mathfrak{T}'$. The sequence $x_n \to u$, and hence $\gamma.x_n$, tends to a point on ∂X̄ in the closure of $\mathfrak{T}'$, which is necessarily equal to v. Hence $\gamma.u = v$.

Remark. These statements are equivalent to the following similar ones for Siegel sets in G. Let $\mathfrak{S}$ and $\mathfrak{T}$ be Siegel sets in G with respect to cuspidal pairs (P, A) and (P', A'), and let C be a relatively compact set in G. Then

$$\{\gamma \in \Gamma \mid \gamma(C) \cap \mathfrak{S} \neq \emptyset\} \quad \text{and} \quad \{\gamma \in \Gamma \mid \gamma(\mathfrak{S}) \cap \mathfrak{T} \neq \emptyset\}$$

are finite. If P' is not conjugate to P under Γ, then $\gamma(\mathfrak{S}) \cap \mathfrak{T}$ is compact and, for sufficiently large t, $\Gamma(\mathfrak{S}_{t,\omega}) \cap \mathfrak{T}_{t',\omega'} = \emptyset$.

This is the form in which these results will usually be employed in the sequel.

3.17 Theorem. *Assume that* Γ *has finite covolume in G. Let* $u_1, \ldots, u_l$ *be a set of representatives of the* Γ-*equivalence classes of cuspidal points for* Γ. *Then* Γ

has a fundamental set $\Omega \subset X$ *of the form* $C \cup (\bigcup_i \mathfrak{S}_i')$, *where* $\mathfrak{S}_i'$ *is a Siegel set at* u_i, *the* $\mathfrak{S}_i'$ *have disjoint images in* $\Gamma \backslash X$, *and* C *is compact. Such a fundamental set has finite area and the Siegel property.*

Proof. Note first that l is finite by 3.14. Since $\Gamma \backslash X_\Gamma^*$ is Hausdorff, we can find disjoint neighborhoods V_i of the u_i. By 3.15(b), we may take for $V_i - \{u_i\}$ the image of a Siegel set $\mathfrak{S}_i'$. Then the $\mathfrak{S}_i'$ are disjoint and have disjoint images in $\Gamma \backslash X$. Since $\Gamma \backslash X_\Gamma^*$ is compact (3.14), the complement of $\bigcup_i V_i$ is relatively compact and is contained in $\Gamma \backslash X$. This complement is the image of a compact subset C of X, so $C \cup (\bigcup_i \mathfrak{S}_i')$ is a fundamental set, which has finite area by 3.15(a). By construction, $\Gamma(\mathfrak{S}_i') \cap \Gamma(\mathfrak{S}_j') = \emptyset$ for $i \neq j$. Therefore the Siegel property follows from 3.16.

Recall that our Siegel sets contain geodesic arcs passing through i. Thus, this fundamental set is a Poincaré one if i is not elliptic for Γ, and is used as the origin in Poincaré's construction.

Remark. By construction, the fundamental domain Ω given by 3.17 is such that $\bar{\Omega} \cap \partial \bar{X}$ is a set of representatives for the Γ-equivalence classes of Γ-cuspidal points. However, the reader should be warned that a fundamental domain of similar type constructed naturally may not satisfy this property. See 3.20.5 for an example.

3.18 The inverse image $\pi^{-1}(\Omega)$ is a fundamental set for Γ in G. It is the union of a compact set $\pi^{-1}(C)$ and of the Siegel sets $\mathfrak{S}_i = \pi^{-1}(\mathfrak{S}_i')$; clearly, it has finite area and the Siegel property. Again, $\Gamma(\mathfrak{S}_i) \cap \Gamma(\mathfrak{S}_j)$ is empty if $i \neq j$.

3.19 Finally, we recall (but will not use in this book) that $\Gamma \backslash X_\Gamma^*$ carries the natural structure of a 1-dimensional complex analytic manifold (see e.g. [34] or [53, 1.3, 1.5]). Here we sketch the construction of the local structure.

If $c \in X$ is not an elliptic point, then $\pi(c)$ has neighborhoods that are isomorphic images of neighborhoods of $c \in X$, whence a complex structure, evidently independent of the choice of c in $\Gamma.c$. Let c be elliptic and k be the order of the isotropy group of x in Γ'. Take $w = z - c$ as local coordinate around c. Two points a, b close to c have the same image in $\Gamma \backslash X_\Gamma^*$ if and only if $w^k(a) = w^k(b)$. Then we may take w^k as a local coordinate around $\pi(c)$.

Let now c be cuspidal. We may assume that $c = \infty$. Then Γ_N is generated by a translation $z \mapsto z + h$ for some $h > 0$. The map $z \mapsto w = \exp(2\pi i z / h)$ defines an isomorphism of $\Gamma_N \backslash H_t$ (where H_t is, as usual, the horostrip $y > t$) onto an open disc D^* in the w-plane centered and punctured at the origin. On the other hand, for t big enough, the equivalence relations on H_t defined by Γ and Γ_N are the same (3.8), so π defines an isomorphism of $\Gamma_N \backslash H_t$ onto a neighborhood U^* of $\pi(c)$ in $\Gamma \backslash X_\Gamma^*$, punctured at $\pi(c)$. We have therefore a homeomorphism of D^* onto U^*. By mapping the origin onto $\pi(c)$, we define a homeomorphism of $D = D^* \cup \{0\}$ onto a neighborhood of $\pi(c)$ in $\Gamma \backslash X_\Gamma^*$, and w provides a local coordinate around $\pi(c)$.

As consequence, if Γ has finite covolume then $\Gamma \backslash X$ may be identified to the complement of finitely many points in a compact Riemann surface.

3.20 Examples of discrete subgroups.   A discrete subgroup of PG is called a *Fuchsian group* (a terminology introduced by H. Poincaré but regarding which F. Klein expressed strong reservations). It is of the *first kind* if it has a fundamental domain of finite area (actually, this is not the original definition of "first kind", but it is equivalent). We next describe some series of Fuchsian groups of the first kind; this is for illustration, to give an idea of which groups we are dealing with. However, the facts to be stated will not be used later, so we shall be freer with the prerequisites.

3.20.1   The most famous Fuchsian group is the modular group, denoted here Γ_1, the image in PG of $SL_2(\mathbb{Z})$. Its standard fundamental domain is the set of $z \in X$ such that $|z| \geqslant 1$ and $|\mathcal{R}z| \leqslant 1/2$, where $-1/2 + iy$ is equivalent to $1/2 + i.y$. The modular group is the free product $\mathbb{Z}/2\mathbb{Z} * \mathbb{Z}/3\mathbb{Z}$ of the groups of order 2 and 3 generated respectively by $w \colon z \mapsto -z^{-1}$ and the rotation of angle $2\pi/3$ at $z = (1 + \sqrt{3})/2$. A *principal congruence subgroup* $\Gamma(n)$ of Γ_1 is the image in PG of the group of elements of $SL_2(\mathbb{Z})$ congruent to the identity modulo n ($n \in \mathbb{N}$, $n > 1$). The quotient $\Gamma_1/\Gamma(n)$ is the finite group $SL_2(\mathbb{Z}/n\mathbb{Z})$ of 2×2 matrices with coefficients in $\mathbb{Z}/n\mathbb{Z}$ and determinant 1. A subgroup Γ of Γ_1 is a *congruence subgroup* if it contains some $\Gamma(n)$. This terminology is also used sometimes for $SL_2(\mathbb{Z})$ and its subgroups.

It has been remarked already by O. Fricke and F. Klein that Γ_1 contains subgroups of finite index that are not congruence subgroups. In fact, any finite group generated by one element of order 2 and one of order 3 is a quotient of Γ_1. Among those are simple groups that are not a homomorphic image of any $SL_2(\mathbb{Z}/n\mathbb{Z})$. The kernel of a homomorphism onto such a group is therefore not a congruence subgroup.

3.20.2   We let T_r ($r \in \mathbb{R}$) be the translation $z \mapsto z + r$. The group Γ_1 is also generated by w and T_1. The group $\Gamma(2)$ has index 6 in Γ_1 and is a free group on two generators. It is contained in the so-called Θ-group Γ_2, generated by w and T_2.

The group Γ_2 has index 3 in Γ_1, has as a fundamental domain a geodesic trian-
gle with the three vertices at infinity given by $|z| \geqslant 1$ and $|\mathcal{R}z| \leqslant 1$, and is the free
product of $\mathbb{Z}/2\mathbb{Z}$ and $\mathbb{Z}$.

The groups Γ_1 and Γ_2 belong to an interesting 1-parameter family of subgroups
Γ_λ ($\lambda \in [1, \infty)$) of PG considered by E. Hecke. The group Γ_λ is generated by w
and T_λ. If $\lambda = 2\cos\pi/m$ ($m \geqslant 3$) then Γ_λ is discrete, isomorphic to $\mathbb{Z}/2\mathbb{Z}*\mathbb{Z}/m\mathbb{Z}$,
and hence a quotient of Γ_2. For other values of $\lambda \in (1, 2)$, Γ_λ is isomorphic to
Γ_2 but not discrete; for $\lambda > 2$ it is discrete, again isomorphic to Γ_2, but not of finite covolume. The phenomenon of a small deformation that is discrete and ho-
momorphic but not isomorphic to the original group also occurs in $SL_2(\mathbb{C})$, as a
consequence of Dehn surgery in hyperbolic 3-manifolds of finite volume; in fact,
the present case can be viewed as a 2-dimensional analog. However, neither of
these three features occurs in general in other semisimple groups, where discrete
subgroups of finite covolume are usually locally rigid (any small deformation is
conjugation by an element of the ambient group).

3.20.3 A geometric construction. Let $I(\mathsf{X})$ be the group of isometries of X
with respect to the Poincaré metric. Its identity component has index 2, is the
group of orientation preserving isometries, and is equal to PG. Given a geodesic
C, the group $I(\mathsf{X})$ contains the *reflection* with respect to C – namely, an invo-
lution leaving C and only C pointwise fixed. (If $C = i\mathbb{R}^{+*}$ then the reflection
is $z \mapsto -\bar{z}$.) Let now F be a geodesic polygon with finitely many sides (vertices at infinity allowed). Assume that the interior angle at a vertex in X is of the
form $2\pi/m$ ($m \geqslant 3$). Then the reflections to the geodesics containing one side of
F can be shown to generate a discrete subgroup with fundamental domain F. Its
intersection with PG is of index 2 and is a Fuchsian group of the first kind.

3.20.4 An arithmetic construction. To complete the picture, let us mention an
arithmetic method of defining cocompact Fuchsian groups. For more details, see
[53, IX].

Let B be a (division) quaternion algebra over $\mathbb{Q}$ that splits at infinity – that is,
such that $B \otimes_\mathbb{Q} \mathbb{R} = M_2(\mathbb{R})$. There are integers $d, q > 1$ such that B is the set of
matrices

$$a = \begin{pmatrix} x & y \\ -q.\bar{y} & \bar{x} \end{pmatrix} \quad \left(x, y \in F = \mathbb{Q}(\sqrt{d}) \right),$$

where $x \mapsto \bar{x}$ is the automorphism $u + v\sqrt{d} \mapsto u - v\sqrt{d}$ ($u, v \in \mathbb{Q}$) of F.
An order in B is a subring that is a finitely generated $\mathbb{Z}$-submodule containing a
$\mathbb{Q}$-basis. Let $\mathfrak{o}$ be a maximal one. Then the group

$$\Gamma = \{a \in B \mid a \in \mathfrak{o}, \det a = 1\}$$

is a cocompact Fuchsian subgroup. More generally, one can start from a totally real field E, a quaternion algebra over E that splits at one real place of E and remains a division algebra at the other real places of E, and define Γ similarly.

There is an overlap between 3.20.3 and 3.20.4. For instance, there are finitely many triangle groups that fall under the heading of 3.20.4 (see [54]).

3.20.5   (i) The closure of the fundamental domain for Γ_2 given in 3.20.2 has three points on $\partial\bar{X}$: ∞, -1, and 1. However, -1 and 1 are equivalent under w, so Γ_2 has at most two cuspidal points up to equivalence and, in fact, does have two.

(ii) The area of $\Gamma\backslash X$ can be expressed as a function of the number l of cusps (mod Γ), of elliptic points and the order of isotropy groups (in Γ') at those points, and of the genus of $\Gamma\backslash X_\Gamma^*$ (see e.g. 2.20 in [53]). From this formula one sees easily that $\mathrm{Ar}(\Gamma_2\backslash X)$ is the minimum of the area among Fuchsian groups with two cusps and that, for $l \geqslant 3$, the minimum of the area among groups with l cusps is realized by the sphere punctured at l points. For $l = 3$ this space is $\Gamma(2)\backslash X$.

4 The unit disc model

The upper half-plane X is conformally equivalent to the open unit disc

$$D = \{ w \in \mathbb{C} \mid |w| < 1 \},$$

and much of Sections 2 and 3 can be given an equivalent description in that framework. We shall mostly stick to X, but the unit disc model will be used in an essential way to prove 5.15 and also in Section 6 to describe Poincaré's argument for the convergence of Poincaré series.

4.1 In the sequel,

$$(1) \qquad T = \begin{pmatrix} 1 & -i \\ 1 & i \end{pmatrix}; \quad \text{hence } T^{-1} = \frac{1}{2}\begin{pmatrix} 1 & 1 \\ i & -i \end{pmatrix} \in GL_2(\mathbb{C}).$$

The transformation

$$(2) \qquad \tau : z \mapsto w = \tau(z) = \frac{z - i}{z + i}$$

maps conformally X onto D and extends continuously to a homeomorphism of $\bar{X}$ onto $\bar{D} = \{ w \in \mathbb{C} \mid |w| \leqslant 1 \}$.

The map $U \mapsto TUT^{-1}$ ($U \in SL_2(\mathbb{C})$) is an automorphism of $SL_2(\mathbb{C})$ that transforms $G = SL_2(\mathbb{R})$ onto

$$(3) \qquad G_D = SU(1, 1) = \left\{ \begin{pmatrix} a & b \\ \bar{b} & \bar{a} \end{pmatrix} \middle| a, b \in \mathbb{C}, |a|^2 - |b|^2 = 1 \right\}.$$

We let also τ denote this isomorphism. Obviously,

$$\tau(g.z) = \tau(g).\tau(z) \quad (g \in G, z \in X)$$

and $G_D / \{\pm 1\}$ is the group of conformal transformations of D. The element $g = \begin{pmatrix} a & b \\ \bar{b} & \bar{a} \end{pmatrix}$ acts by

$$(4) \qquad w \mapsto \frac{aw + b}{\bar{b}w + \bar{a}}.$$

Again,

$$(5) \qquad \frac{dg}{dw} = (\bar{b}w + \bar{a})^{-2},$$

and the canonical automorphy factor is

$$(6) \qquad \mu_D(g, w) = \bar{b}w + \bar{a}.$$

We shall write $w = u + iv$ ($u, v \in \mathbb{R}$). The invariant metric and volume elements are, respectively,

$$(7) \qquad \frac{du^2 + dv^2}{(1 - |w|^2)^2} \quad \text{and} \quad \frac{du \wedge dv}{(1 - |w|^2)^2}.$$

48

4.2 The map τ sends i to the origin and K to

(1) $$K_D = \left\{ \begin{pmatrix} a & 0 \\ 0 & \bar{a} \end{pmatrix} \,\middle|\, |a| = 1 \right\}.$$

More precisely,

(2) $$T \begin{pmatrix} \cos\varphi & \sin\varphi \\ -\sin\varphi & \cos\varphi \end{pmatrix} T^{-1} = \begin{pmatrix} e^{i\varphi} & 0 \\ 0 & e^{-i\varphi} \end{pmatrix}.$$

Again identify $\mathbb{R}/2\pi$ with K_D by the rule

$$\varphi \mapsto k_\varphi = \begin{pmatrix} e^{i\varphi} & 0 \\ 0 & e^{-i\varphi} \end{pmatrix}$$

and let χ_m be the character of K_D defined by $\chi_m(k_\varphi) = e^{im\varphi}$. Then, if f is a function on G of type m on the right, the transform $\tilde{f}: g \mapsto f(\tau^{-1}(g))$ on G_D is of type m on the right for K_D.

Note that

(3) $$\mu_D(k, 0) = \chi_{-1}(k) \quad (k \in K_D).$$

Moreover, since $K_D = \exp \mathbb{R}iH$, a function f on G_D is of right K_D-type m if and only if $iHf = imf$ (same argument as in 2.12), hence if and only if

(4) $$Hf = mf.$$

4.3 We view D as embedded on the Riemann sphere $\mathbb{C} \cup \infty$ on which $SL_2(\mathbb{C})$ acts, also by fractional linear transformations. The stability group of zero in $SL_2(\mathbb{C})$ is

(1) $$B^- = \left\{ \begin{pmatrix} a & 0 \\ b & a^{-1} \end{pmatrix} \,\middle|\, a, b \in \mathbb{C}, \, a \neq 0 \right\},$$

the intersection of which with G_D is, of course, K_D. Let

(2) $$N_c = \exp \mathbb{C}E, \quad N_c^- = \exp \mathbb{C}F, \quad K_{D,c} = \exp \mathbb{C}H.$$

The Lie algebra $\mathfrak{g}$ of G_D is spanned by

(3) $$iH, \, E + F, \, i(E - F),$$

where $E + F$ and $i(E - F)$ span over $\mathbb{R}$ the (real) tangent space $T(D)_0$ to D at zero. The complexification of this space is spanned by E and F. Moreover,

$$w \mapsto e^{wE}(0) = w$$

yields an isomorphism of $|w| < 1$ onto D. Hence E spans the holomorphic part $T(D)_{0,c}^{1,0}$ and F the antiholomorphic part $T(D)_{0,c}^{0,1}$ of the complexified tangent space to D at 0. If, in analogy with 2.5, we denote by Y^* the differential operator on the Riemann sphere defined by $Y \in \mathfrak{g}_c$, we have

$$(4) \qquad\qquad E_0^* = \frac{\partial}{\partial z} \quad \text{and} \quad F_0^* = \frac{\partial}{\partial \bar{z}}.$$

In Section 5 we shall need the following lemma.

4.4 Lemma. *Let* $g_+ \in N_c$, $g_0 \in K_D$, *and* $g_- \in N_c^-$. *Then*

$$(1) \qquad\qquad \mu_D(g_+ \cdot g_0 \cdot g_-, 0) = \chi_{-1}(g_0).$$

Proof. Since g_0 and g_- fix the origin, the cocycle identity yields

$$\mu_D(g_+ \cdot g_0 \cdot g_-, 0) = \mu_D(g_+, 0) \cdot \mu_D(g_0 \cdot g_-, 0)$$
$$= \mu_D(g_+, 0) \cdot \mu_D(g_0, 0) \cdot \mu_D(g_-, 0).$$

Hence the lemma follows from 4.2(3) and the obvious relations

$$\mu_D(g_+, 0) = \mu_D(g_-, 0) = 1.$$

In Section 6, we shall use the following lemma.

4.5 Lemma. *Let* $m \in \mathbb{Z}$. *The function* $g \mapsto \mu_D(g, 0)^{-m}$ *is bounded on G if* $m \geqslant 0$, *and belongs to* $L^2(G)$ *if* $m \geqslant 2$ *and to* $L^1(G)$ *if* $m \geqslant 4$.

Proof. Write $g = \begin{pmatrix} a & b \\ \bar{b} & \bar{a} \end{pmatrix}$. Then

$$(1) \qquad\qquad |a|^2 - |b|^2 = 1$$

as already recalled, and

$$(2) \qquad\qquad \mu_D(g, 0) = \bar{a}.$$

From (1), we see that $|a|^2 \geqslant 1$, hence also $|a| \geqslant 1$ and therefore

$$(3) \qquad\qquad |a|^{-m} = |\mu_D(g, 0)|^{-m} \leqslant 1 \quad (m \in \mathbb{N}),$$

which proves the first assertion.

Let $w = g.0$. We have

$$(4) \qquad\qquad w = b.\bar{a}^{-1}, \quad \frac{dg}{dw}(0) = \bar{a}^{-2}.$$

Hence, in view of (1),

$$(5) \qquad\qquad |a|^2(1 - |w|^2) = 1,$$

and the invariant volume element dv on D in 4.1(7) is equal to

$$(6) \qquad dv_w = \left| \frac{dg}{dw}(0) \right|^{-2} du \wedge dv = |a|^4 \, du \wedge dv.$$

The function $g \mapsto |\mu_D(g, 0)|$ is K-invariant on the right and on the left, as follows from the cocycle identity and the fact that $|\mu_D(k, z)| = 1$ for every $z \in D$ and $k \in K_D$. From the invariance on the right and (2) we have

$$(7) \qquad \int_G |\mu_D(g, 0)|^{-m} \, dg = c \int_D |\mu_D(g, 0)|^{-m} |a|^4 \, du \wedge dv = c \int_D |a|^{4-m} \, du \wedge dv,$$

and (3) implies that the last integral is finite for $m \geqslant 4$. For square integrability, we need $\int_G |\mu_D(g, 0)|^{-2m} \, dg < \infty$, hence $m \geqslant 2$.

5 Growth conditions. Automorphic forms

5.1 Growth conditions. Let ∞ be cuspidal for Γ and let $\Gamma_\infty = \Gamma \cap N_0$. Let f be a function on X that is Γ_∞-invariant on the left and of right K-type $\chi \in \hat{K}$, that is,

(1) $$f(\gamma g k) = f(g)\chi(k) \quad (\gamma \in \Gamma_\infty,\ g \in G,\ k \in K).$$

Let (P, A) be a (normal, cf. 2.6) p-pair, and let $\mathfrak{S} = \omega.A_t.K$ be a Siegel set with respect to (P, A) (see 3.15). Then f is said to be of *moderate growth* on $\mathfrak{S}$ if there exists $\lambda \in \mathbb{R}$ such that

(2) $$|f(g)| \prec a(g)^\lambda \quad (g \in \mathfrak{S}).$$

If f is smooth then it is of *uniform moderate growth* if

(3) $$|Df(g)| \prec a(g)^\lambda \quad (g \in \mathfrak{S},\ D \in \mathcal{U}(\mathfrak{g}));$$

f is *rapidly decreasing* if (2) is true for all λ and *uniformly rapidly decreasing* if (3) holds for all λ and all D.

Remark. Let (P', A') be a second p-pair. There exists $k \in K$ that conjugates (P', A') to (P, A). If f satisfies any of these conditions, then $l_k f$ satisfies the analogous condition on a Siegel set with respect to (P', A').

Assume now that F is a holomorphic function that is Γ_∞-invariant and admits a convergent power series development

(4) $$F(z) = \sum_{n \geqslant n_0} a_n e^{2\pi i n z/c}$$

where c is such that $\langle T_c \rangle = \Gamma_\infty$. Since $|e^{2\pi i n z/c}| = e^{-2\pi n y/c}$, we see that

$$F \text{ has moderate growth} \Longleftrightarrow a_n = 0,\ n < 0,$$
$$F \text{ is rapidly decreasing} \Longleftrightarrow a_n = 0,\ n \leqslant 0.$$

Moreover, F is fast increasing (i.e., increasing faster than any power of y) if $a_n \neq 0$ for some $n < 0$.

5.2 Proposition. *Assume that f has moderate growth on $\mathfrak{S}_t$ with exponent λ. Then so has $f * \alpha$ for any $\alpha \in C_c^\infty(G)$.*

Proof. We claim first that if $U \subset G$ is relatively compact and $x \in NA_tK$, then there exists $t' > 0$ such that

(1) $$xU \subset NA_{t'}K.$$

Since KU is relatively compact, there exists $C_A \subset A$ and $C_N \subset N$ compact such that
$$KU \subset C_N C_A K.$$

Clearly, $aC_N \subset NA$ if $a \in A$ since N is normal in NA. Hence
$$xU = na(x)kU \subset Na(x)C_N C_A K \subset Na(x)C_A K.$$

Obviously, $a^\lambda \asymp 1$ on C_A for any λ. Since $(a(x)a)^\lambda = a(x)^\lambda.a^\lambda$, the assertion (1) follows.

Assume now that $\operatorname{supp}(\alpha) \subset U$ and $U = U^{-1}$. We have

(2) $$|f * \alpha(x)| = \left| \int_G f(y)\alpha(y^{-1}.x)\,dy \right| \leqslant \|\alpha\|_\infty \int_{xU} |f(y)|\,dy.$$

If $|f(x)| \prec a(x)^\lambda$ on $NA_{t'}K$ we have, using (1) and (2),
$$|(f * \alpha)(x)| \prec \|\alpha\|_\infty \operatorname{vol}(U)a(x)^\lambda \quad \text{for } x \in NA_tK,$$

whence the proposition.

5.3 Corollary. *Assume that $g = f * \alpha$ ($\alpha \in C_c^\infty(G)$). Then*
$$|Dg(x)| \prec a(x)^\lambda \quad \text{for any } D \in \mathcal{U}(\mathfrak{g}) \quad (x \in \mathfrak{S}_{t,\omega}).$$

This follows from the proposition and the equalities

(1) $$Dg(x) = D(f * \alpha)(x) = (f * D\alpha)(x).$$

5.4 Classical case. If f satisfies 5.1(1), then $|f|$ may be viewed as a function on X. In particular, let $(P, A) = (P_0, A_0)$ and let $\mathfrak{S}_0$ be a standard Siegel set $|x| \leqslant c$, $y \geqslant t$. Then 5.1(2) can be written as
$$|f(z)| \prec y^\lambda \quad (z \in \mathfrak{S}_0).$$

5.5 Definition. A smooth function $f: G \to \mathbb{C}$ is an *automorphic form* for Γ if it satisfies the following conditions:

A1 $f(\gamma g) = f(g)$ ($\gamma \in \Gamma$, $g \in G$);
A2 f is K-finite on the right;
A3 f is Z-finite – that is, f is annihilated by an ideal J of finite codimension in $\mathcal{Z}(\mathfrak{g}) = \mathcal{Z}$;
A4 f is of moderate growth at the cusps of Γ.

Remarks. (1) Here, $\mathcal{Z} = \mathbb{C}[\mathcal{C}]$, hence A3 is equivalent to the existence of a non-constant polynomial $P(\mathcal{C})$ in the Casimir operator which annihilates f. The ideal J is generated by such a polynomial of smallest possible degree.

(2) If f is K-finite on the right then it is a finite sum of functions f_i, where f_i has the right K-type n_i for some $n_i \in \mathbb{Z}$; it would therefore not be an essential restriction to assume in A2 that f is of type $\chi \in \hat{K}$.

The space of automorphic forms of type χ, J is denoted by $\mathcal{A}(\Gamma, J, \chi)$. We shall see later that this space is finite dimensional.

5.6 First properties of automorphic forms. Let f be an automorphic form. Then

(a) f is real analytic.

This follows from 2.13, in view of A2 and A3.

(b) Given any neighborhood U of 1, there exists $\alpha \in I_c^\infty(G)$ with supp $\alpha \subset U$ such that

$$f = f * \alpha.$$

This is a consequence of A2, A3, and 2.14.

(c) f has uniform moderate growth at every cusp of Γ.

This follows from 5.3 and (b).

5.7 Proposition. *Let $p \geqslant 1$ and $\varphi \in C_c(G)$, and let P be a cuspidal parabolic subgroup for Γ. Then there exists a constant $c_\varphi > 0$ such that*

(1) $|(f * \varphi)(x)| \leqslant c_\varphi \|f\|_p a(x)^{2\rho/p} \quad (x \in NA_t K)$

for all $f \in L^p(\Gamma \backslash G)$.

Proof. The function f is Γ_N-invariant on the left, so we may assume x to be contained in the Siegel set

$$\mathfrak{S}_{P,t} = \omega A_t K,$$

where ω is a relatively compact subset of N containing a fundamental domain for $\Gamma_N = N \cap \Gamma$ (3.15).

Let C be a compact set containing supp (φ). Then, as in 5.2, we have

$$KC \subseteq C_N C_A K$$

with $C_A \subset A$ and $C_N \subset N$ compact, hence

$$xC \subseteq \omega a(x)KC \subseteq \omega a(x)C_N C_A K$$
$$\subseteq a(x)\big(a(x)^{-1}\omega a(x)C_N\big)a(x)^{-1}a(x)C_A K.$$

By definition, we have

$$(f * \varphi)(x) = \int_G f(y)\varphi(y^{-1}x)\, dy,$$

whence

$$|f * \varphi(x)| \leqslant \|\varphi_\infty\| \int_{xC} |f(y)|\, dy.$$

Since $a(x)^{-2\rho} \prec 1$ on $\mathfrak{S}_{P,t}$, the set $a(x)^{-1}\omega a(x)C_N \subset N$ is bounded and hence contained in a compact set D. Therefore

$$xC \subset a(x)Da(x^{-1})a(x)C_A K.$$

The automorphism $n \mapsto a(x)na(x)^{-1}$ of $N \cong \mathbb{R}$ is the dilation by $a(x)^{2\rho}$. Therefore

$$\ell\big(a(x)Da(x)^{-1}\big) \leqslant a(x)^{2\rho}\ell(D),$$

where ℓ is the ordinary measure on $\mathbb{R}$. Let $Q(x) = D.a(x).C_A.K$. Because $|f(y)|$ is left-invariant under Γ_N, we can write

$$\int_{xC} |f(y)|\, dy \leqslant a(x)^{2\rho} \int_{Q(x)} |f(y)|\, dy.$$

Let q be the conjugate exponent to p, that is, $p^{-1} + q^{-1} = 1$. By Hölder's inequality,

$$\int_{Q(x)} |f(y)|\, dy \leqslant \left(\int_{Q(x)} |f(y)|^p\, dy \right)^{1/p} \left(\int_{Q(x)} dy \right)^{1/q}.$$

The first integral on the right-hand side is bounded by $\|f\|_p$; the second integral is $\asymp a(x)^{-2\rho/q}$ (use 2.9(4)). Altogether, we have

$$\int_{Q(x)} |f(y)|\, dy \leqslant \delta.\|f\|_p a(x)^{2\rho - 2\rho/q} = \delta.\|f\|_p.a(x)^{2\rho/p},$$

with δ independent of f, and the proposition follows.

5.8 Corollary. *If f satisfies conditions A1–A3 in 5.5 and belongs to $L^p(\Gamma \backslash G)$ for some $p \geqslant 1$, then f satisfies A4 in 5.5 (i.e., f is an automorphic form).*

5.9 The group Γ has a fundamental domain that is the union of a compact set and finitely many disjoint Siegel sets $\mathfrak{S}_1, \dots, \mathfrak{S}_l$ (3.18). We want to show that it suffices to check the growth condition at the $\mathfrak{S}_i$ by giving an intrinsic definition of moderate growth not involving Siegel sets. We let

$$(1) \qquad\qquad \|x\| = (\mathrm{tr}({}^t x.x))^{1/2} \quad (x \in G)$$

be the Hilbert–Schmidt norm of x. Thus

(2) $$\|x\|^2 = a^2 + b^2 + c^2 + d^2 \quad \text{if } x = \begin{pmatrix} a & b \\ c & d \end{pmatrix}.$$

The following properties are elementary:

(3)
$$\|x\| \geqslant 1, \quad \|x\| = \|x\|^{-1}, \quad \|x.y\| \leqslant \|x\| \cdot \|y\|, \quad \|k\|^2 = 2$$
$$(x, y \in G, \, k \in K).$$

Given $C \subset G$ compact,

(4) $$\|y.x.z\| \asymp \|x\| \quad (x \in G, \, y, z \in C).$$

Now let (P, A) be a p-pair and let $\mathfrak{S}_t$ be a Siegel set with respect to (P, A). Then

(5) $$\|x\| \asymp \|a(x)\| \asymp a(x)^\rho = h_P(x)$$

(notation of 2.8). The first relation follows from (4), the second from the fact that, up to conjugation by a fixed element of K, the element $a(x)$ is the diagonal matrix with entries $a(x)^\rho$ and $a(x)^{-\rho}$, and the third one from the definition. As a consequence, for a function f on $\mathfrak{S}_t$ we have:

(6)   there exists $m \in \mathbb{N}$ such that $|f(x)| \prec \|x\|^m \Longleftrightarrow f$ has moderate growth;

(7) $|f(x)| \prec \|x\|^m$ for all $m \in \mathbb{Z} \Longleftrightarrow f$ is fast decreasing.

5.10 Proposition. *Assume that Γ has finite covolume, and let Ω be a fundamental set for Γ in G as in 3.18. Then*

(1) $$\|\gamma.x\| \succ \|x\| \quad (\gamma \in \Gamma, \, x \in \Omega).$$

Proof. If x varies in a compact set, then $\|\gamma.x\| \asymp \|\gamma\|$. Since $\|\gamma\| \geqslant 1$, this proves (1) for x in a compact subset of Ω. It suffices then to prove (1) when x varies in one of the $\mathfrak{S}_i$. Let $k \in K$. If the assertion is proved for $^k\Gamma$ and $^k\mathfrak{S}_i$ then, in view of 5.9(4), it holds also for Γ and $\mathfrak{S}_i$. Changing the notation and replacing Γ by a conjugate if necessary, we may assume that ∞ is cuspidal for Γ and that $\mathfrak{S}_i$ is a Siegel set consisting of elements

(2) $$x = \begin{pmatrix} 1 & u \\ 0 & 1 \end{pmatrix} \cdot \begin{pmatrix} v & 0 \\ 0 & v^{-1} \end{pmatrix} .k. \quad (|u| \leqslant c, \, v > t, \, k \in K),$$

where c and t are strictly positive constants. Thus

(3) $$x = \begin{pmatrix} v & uv^{-1} \\ 0 & v^{-1} \end{pmatrix} .k$$

and so

(4) $$\|x\| \asymp |v| \quad (x \in \mathfrak{S}_i),$$

as we have already pointed out. Let now $\gamma \in \Gamma_\infty$. Then γ is upper triangular, with ± 1 in the diagonal. The first entry of $\gamma.x.k^{-1}$ is $\pm v$, hence $\|\gamma x\| \succ \|x\|$ by (2). Let now $\gamma \in \Gamma - \Gamma_\infty$, with γ_3 the lower left entry of γ; then that of $\gamma.x$ is $\gamma_3.v$. By the lemma in 3.7, there exists a strictly positive constant r such that $|\gamma_3| \geqslant r$ for all $\gamma \in \Gamma - \Gamma_\infty$. Therefore $\|\gamma.x\| > r.|v|$, and the lemma follows.

Remark. The relation (1) can also be written

(2) $$\min_{\gamma \in \Gamma} \|\gamma.x\| \asymp \|x\| \quad \text{for } x \in \Omega.$$

5.11 Proposition. *We keep the previous notation. Let f be a function on $\Gamma \backslash G$. The following conditions are equivalent:*

(i) *there exists $m \in \mathbb{N}$ such that $|f(x)| \prec \|x\|^m$ $(x \in G)$;*
(ii) *f is of moderate growth on all the cusps for Γ;*
(iii) *f has moderate growth on the $\mathfrak{S}_i$ $(1 \leqslant i \leqslant l)$.*

Proof. The implication (i) $\Rightarrow$ (ii) follows from 5.9(6), and (ii) $\Rightarrow$ (iii) is obvious. Assume now (iii). Then, by 5.9(6), there exists $m \in \mathbb{N}$ such that

$$|f(x)| \prec \|x\|^m \quad (x \in \Omega).$$

By 5.10, there exists a constant $c > 0$ such that $\|x\|^m \leqslant c.\|\gamma.x\|^m$ for all $\gamma \in \Gamma$ and $x \in \Omega$. Since $f(\gamma.x) = f(x)$ $(\gamma \in \Gamma, x \in G)$, condition (i) holds.

Remark. From 5.9(7) we see that a function f on $\Gamma \backslash G$ is fast decreasing on $\mathfrak{S}_i$ if and only if $|f(x)| \prec \|x\|^m$ for all $m \in \mathbb{Z}$.

5.12 Lemma. (i) *For $t > 0$ let $G_t = \{ x \in G \mid \|x\| \leqslant t \}$. Then there exists a constant $c > 0$ such that*

(1) $$\mathrm{vol}(G_t) \leqslant c.t^3 \quad (t > 0).$$

(ii) *Let Σ be a discrete subgroup of G. Then there exists a constant $d > 0$ such that*

(2) $$\mathrm{Card}(G_t \cap \Sigma) \leqslant d.t^3 \quad (t > 0).$$

Proof. (i) We use the standard Iwasawa decomposition $G = N.A.K$ with respect to the upper triangular group. By 5.9(4),

$$\|xk\| \asymp \|x\| \quad (x \in G, k \in K),$$

so it suffices to prove the existence of $c > 0$ such that

$$\int_{(NA)_t} a^{-2\rho} \, da.dn \leqslant c.t^3,$$

where $(NA)_t = G_t \cap NA$ (cf. 2.9 for the measure). Note that G_t is empty for $t < 1$, so we assume $t \geqslant 1$. If $x = \begin{pmatrix} q & n \\ 0 & q^{-1} \end{pmatrix}$ then $\|x\| \leqslant t$ implies $q \in [t^{-1}, t]$ and $|n| \leqslant t$.

The measure on A can be written $q^{-3}.dq$ and the measure on N is the Euclidean one, so we obtain

$$\text{vol}(G_t) \leqslant c' \int_{t^{-1}}^{t} \frac{dq}{q^3} \int_{-t}^{t} dn = c'(t^2 + t^{-2})t$$

for some constant $c' > 0$, and (i) follows.

(ii) There exists a compact symmetric neighborhood U of 1 such that

$$\Sigma \cap U.U = \{1\}.$$

Then the translates $\sigma.U$ ($\sigma \in \Sigma$) are disjoint. Let r be the maximum of $\|\cdot\|$ on U. If $\sigma \in G_t$ then $\sigma.U \subset G_{t.r}$, so

$$\text{Card}(\Sigma \cap G_t).\text{vol}(U) \leqslant \text{vol } G_{t.r}$$

and hence (ii) follows from (i).

We now want to relate our notion of automorphic form to the classical one of holomorphic automorphic form of weight m. We first discuss an easy formalism that allows us to go from functions on X to functions on G.

5.13 Lemma. *Fix $m \in \mathbb{Z}$.*
(i) *Let f be a function on X, and define $\tilde{f}: G \to \mathbb{C}$ by*

(1) $$\tilde{f}(x) = f(x.i).\mu(x, i)^{-m}.$$

Then $\tilde{f}$ is of right K-type m.
(ii) *Let G operate on functions on X by the rule*

$$(x \circ f)(z) = f(x.z)\mu(x, z)^{-m} \quad (z \in \mathsf{X}, \ x \in G).$$

Then $(x^{-1} \circ f)^{\sim} = l_x \tilde{f}$.

Proof. (i) K is the isotropy group of i, so the map $k \mapsto \mu(k, i)$ is a character of K. In fact, if

$$k = k_\theta = \begin{pmatrix} \cos\theta & \sin\theta \\ -\sin\theta & \cos\theta \end{pmatrix}$$

then $\mu(k, i) = e^{-i\theta}$; that is,

(2) $$\mu(k, i) = \chi_{-1}(k).$$

Thus, by the cocycle identity, we have

$$\tilde{f}(x.k) = f(x.k.i).\mu(x.k, i)^{-m}$$
$$= f(x.i).\mu(x, i)^{-m}\mu(k, i)^{-m} = \tilde{f}(x).\chi_m(k) \quad (x \in G, \, k \in K).$$

(ii) By definition and the cocycle identity we have, for $y \in G$,

$$(x^{-1} \circ f)^\sim(y) = (x^{-1} \circ f)(y.i).\mu(y, i)^{-m},$$
$$(x^{-1} \circ f)^\sim(y) = f(x^{-1}.y.i)\mu(x^{-1}, y.i)^{-m}\mu(y, i)^{-m},$$
$$(x^{-1} \circ f)^\sim(y) = f(x^{-1}.y.i)\mu(x^{-1}.y, i)^{-m} = (l_x\tilde{f})(y).$$

5.14 Classical automorphic forms. A holomorphic function $f: X \to \mathbb{C}$ is an automorphic form for Γ of weight $m \in \mathbb{N}$ if

(i) $f(\gamma z) = \mu(\gamma, z)^m f(z)$ $(\gamma \in \Gamma, \, z \in X)$;
(ii) f is regular at the cusps: if $p \in \partial\bar{X}$ is a cuspidal point for Γ and $g \in K$ is such that $g.p = \infty$, then the function $f_g: z \mapsto f(g.z)$ admits a convergent power series development

$$f_g(z) = \sum_{n \geq 0} c_n.e^{2\pi i n z/c},$$

where c is such that $^g\Gamma \cap N = \langle T_c \rangle$.

We want to relate this notion to that of automorphic form in the sense of 5.5. For this, we associate to f the function $\tilde{f}: G \to \mathbb{C}$ defined by

(1) $$\tilde{f}(g) = f(g.i)\mu(g, i)^{-m} \quad (g \in G)$$

and claim that $\tilde{f}$ is an automorphic form.

In the notation of 5.13(ii), the condition of (i) is equivalent to $(\gamma \circ f) = f$ $(\gamma \in \Gamma)$. By 5.13(ii), this yields A1. Furthermore, $\tilde{f}$ is of right K-type m by 5.13, so A2 is fulfilled. We now check the growth condition A4. In view of the remark in 5.1, it suffices to verify it for the standard cusp ∞, assuming it is one for Γ.

We already pointed out in 5.1 that if F satisfies (ii) then $g \mapsto F(g.i)$ is of moderate growth on a standard Siegel set $\omega A_{0,t} K$. There remains to check that $\mu(g, i)^{-m}$ is also of moderate growth. By the cocycle identity,

$$\mu(nak, i) = \mu(n, a.i).\mu(a, i)\mu(k, i).$$

But $\mu(n, z) \equiv 1$ for $n \in N_0$, $\mu(k, i)$ is of modulus 1, and $\mu(a, i) = t_a^{-1} = a^{-\rho}$; hence

$$\mu(g, i)^{-m} = a(g)^{\rho m}$$

is of moderate growth.

Finally, A3 is a consequence of the following, more precise, lemma.

5.15 Lemma. *Let f be a holomorphic function on X, and let $\tilde{f}: G \to \mathbb{C}$ be defined by 5.13(1). Then $C\tilde{f} = (m^2/2 - m)\tilde{f}$.*

To prove this we use the unit disc model and go over to the framework of Section 4, the notation of which we use freely. Using τ, we transport f and $\tilde{f}$ to functions r and $\tilde{r}$ on D and G_D respectively. In view of 5.7(1), we have

(1) $\tilde{r}(g) = r(g.0)\mu_D(g, 0)^{-m}$ $(g \in G_D)$

and that $\tilde{r}$ is of type m on the right for K_D. We claim first that

(2) $Fr(g) = 0$ for all $g \in G_D$.

Assume it provisorily. Then

$$C\tilde{r} = \left(\frac{H^2}{2} - H \right) \tilde{r}$$

and the lemma follows from 4.2(4). There remains to prove (2).

Let $Y \in \mathfrak{g}_c$ be of the form

$$Y = Y_+ + Y_0 + Y_-   (Y_+ \in \mathbb{C}E, \ Y_0 \in \mathbb{R}iH, \ Y_- \in \mathbb{C}F).$$

Then, for t small enough, we have

$$e^{tY} = a_+(t).k(t).a_-(t)   \left(a_+(t) \in N_c, \ a_-(t) \in N_c^-, \ k(t) \in K_D \right)$$

and, clearly,

(3) $Y_0 = \left. \dfrac{dk(t)}{dt} \right|_{t=0}$.

Using (1) and 4.2(3), we obtain

$$\tilde{r}(e^{tY}) = \mu_D(e^{tY}, 0)^{-m} r(e^{tY}.0) = \chi_m(k_0(t)).r(e^{tY}.0);$$

hence, by (3),

$$Y\tilde{r}(1) = d\chi_m(Y_0)r(0) + Y_0^* r(0).$$

Let $Y = F$; then $Y_0 = 0$. Since $F_0^* = \frac{\partial}{\partial \bar{z}}$ and R is holomorphic, we derive (2) for $g = 1$. Given $g \in G_D$, consider the left translate $l_g\tilde{r}$ of $\tilde{r}$, that is,

$$l_g \tilde{r}(x) = \tilde{r}(g^{-1}x) \quad (x \in G_D).$$

Using the cocycle identity we get

$$l_g \tilde{r}(x) = \mu(g^{-1}, x.0)^{-m} \mu(x, 0)^{-m} r(g^{-1}.x.0).$$

Set $w = x.0$. The function

$$w \mapsto \mu(g^{-1}, w)^{-m} r(g^{-1}.w) = r'(w)$$

is holomorphic, and $l_g \tilde{r}(x) = \mu(x, 0)^{-m} r'(x.0)$. The previous argument, applied to the pair $(l_g \tilde{r}, r')$, shows that $Fl_g \tilde{r}(1) = 0$. But F is left-invariant, whence $Fl_g r = l_g Fr$ and

$$l_g F\tilde{r}(1) = F\tilde{r}(g^{-1}) = 0;$$

this proves (2).

5.16 Similar considerations are valid for antiholomorphic functions. They can be proved by essentially the same computations, or by reduction to the holomorphic case by going over to complex conjugates. We leave the details to the reader.

An antiholomorphic function f on X is an automorphic form of weight m if $\bar{f}$ is. This means that $\mu(\gamma, z)$ is replaced by $\overline{\mu(\gamma, z)}$ in 5.14(i) and z by $\bar{z}$ in 5.14(ii). If f is defined by

$$(1) \qquad\qquad \tilde{f}(x) = f(x.i)\overline{\mu(x, i)}^{-m}$$

then $\tilde{f}$ is an automorphic form of right K-type $-m$. More generally, if f is antiholomorphic and $\tilde{f}$ is defined by (1), then $\tilde{f}$ is of right K-type $-m$.

In the following lemma we again switch to the unit disc model D. As before, w is the coordinate on D.

5.17 Lemma. *Let* $m \in \mathbb{Z}$, $n \in \mathbb{N}$, *and* $f_n = w^n$. *Let* f *be defined by*

$$\tilde{f}_n(x) = f_n(x.0).\mu(x, 0)^{-m}.$$

Then $\tilde{f}_n$ *(resp.* $\tilde{\bar{f}}_n$*) has right K-type m (resp. $-m$) and left K-type $m + 2n$ (resp. $-m - 2n$). Let f be holomorphic (resp. antiholomorphic) on* D. *Then* $\tilde{f}$ *is left K-finite if and only if f is a polynomial in w (resp. $\bar{w}$).*

Proof. It suffices to show 5.17 in the holomorphic case. The assertion on the right K-type has been proved earlier. On D, the group K operates by dilations; in fact,

$$(1) \qquad\qquad k.w = \chi_2(k).w \quad (k \in K, \, w \in D).$$

Note also that

$$(2) \qquad\qquad \mu(k, w) = \chi_{-1}(k).w \quad (k \in K),$$

as follows from the definitions. Then

$$\tilde{f}_n(k.x) = f_n(k.x.0)\mu(k.x, 0)^{-m} = \chi_{2n}(k)f_n(x.0).\mu(k.x, 0)^{-m}\mu(k, x.0)^{-m},$$

(3) $$\qquad\qquad \tilde{f}_n(k.x) = \chi_{2n+m}(k)\tilde{f}_n(x).$$

Let now f be holomorphic and assume that $\tilde{f}$ is left K-finite. It follows from 5.13(ii) that $\tilde{f}$ can be written as a finite sum of functions $\tilde{f}_j$ with f_j holomorphic, which are of some given K-type. Hence we may assume that $l_k\tilde{f} = \chi_q(k)\tilde{f}$ for some $q \in \mathbb{Z}$ and all $k \in K$. Again by 5.13, this translates to

(1) $$\qquad\qquad (k \circ f)(w) = \chi_{-q}(k).f(w) \quad (k \in K, \ w \in D),$$

(2) $$\qquad\qquad \chi_{-q}(k)f(w) = f(kw).\mu(k, w)^{-m} \quad (k \in K, \ w \in W).$$

We can write f as a convergent power series

$$f(w) = \sum_{n \geqslant 0} c_n.w^n$$

and obtain

$$\chi_{-q}(k)\sum c_n w^n = \chi_m(k)\sum c_n\chi_{2n}(k)w^n \quad (k \in K, \ w \in D).$$

This implies that $-q = m + 2j$ for some $j \in \mathbb{N}$ and that $c_n = 0$ for $n \neq j$.

5.18 Petersson product. Using 5.13(1) we obtain a simple interpretation of the Petersson scalar product:

(1) $$\qquad\qquad \langle f, f' \rangle = \int_{\Gamma\backslash X} y^{m-2} f(z)\overline{f'(z)}\, dx\, dy,$$

where f, f' are of weight m. Consider the functions $\tilde{f}, \tilde{f}'$ on $\Gamma\backslash G$ associated to f and f' by 5.13(1):

$$\tilde{f}(x) = f(x.i)\mu(x, i)^{-m} \quad \text{and} \quad \tilde{f}'(x) = f'(x.i)\mu(x, i)^{-m} \quad (x \in G).$$

We claim that the Petersson product (1) is just the usual scalar product

(2) $$\qquad\qquad (\tilde{f}, \tilde{f}') = \int_{\Gamma\backslash G} \tilde{f}(x)\overline{\tilde{f}'(x)}\, dx.$$

To see this, note that

$$\tilde{f}(x)\overline{\tilde{f}'(x)} = f(x.i)\overline{f'(x.i)}|\mu(x, i)|^{-2m}.$$

From

$$|\mu(x, i)|^2 = c^2 + d^2 = (\mathcal{I}x(i))^{-1}$$

(see 1.1(6)) we have

$$(\tilde{f}, \tilde{f}') = \int_{\Gamma \backslash G} f(x.i) \overline{f'(xi)} (\mathcal{I}x(i))^m \, dx.$$

Write now z for $x.i$ and y for $\mathcal{I}(x.i)$. The integrand is right-invariant under K, which leaves i fixed; hence the integral can be written

$$(\tilde{f}, \tilde{f}') = \int_{\Gamma \backslash X} f(x.i) \overline{f'(x.i)} y^{m-2} \, dx \, dy,$$

since $y^{-2} \, dx \, dy = dr$ is the invariant measure on X and $dx = dk \, dr$. In this computation, Γ could be any discrete subgroup of G.

6 Poincaré series

6.1 Theorem. *Let φ be a function on G that is integrable and $\mathcal{Z}$-finite. Let p_φ be the sum of the series*

$$\tag{1} p_\varphi(x) = \sum_{\gamma \in \Gamma} \varphi(\gamma x).$$

(i) *If φ is K-finite on the right, then the series p_φ converges absolutely and locally uniformly, belongs to $L^1(\Gamma \backslash G)$, and represents an automorphic form for Γ.*

(ii) *If φ is K-finite on the left, then p_φ converges absolutely and is bounded on G.*

Proof. (i) In view of the assumption and 2.14, there exists $\alpha \in I_c^\infty(G)$ such that $\varphi = \varphi * \alpha$. Fix a relatively compact open neighborhood U of 1 that is symmetric ($U = U^{-1}$) and contains supp α. The relation

$$\tag{2} \varphi(\gamma x) = \int_G \varphi(\gamma x y)\alpha(y^{-1})\,dy = \int_U \varphi(\gamma x y)\alpha(y^{-1})\,dy$$

yields

$$\tag{3} |\varphi(\gamma x)| \leqslant \|\alpha\|_\infty \int_{\gamma x U} |\varphi(y)|\,dy.$$

Fix a compact subset C of G. We want to prove absolute and uniform convergence on C. Since $C.U$ is relatively compact,

$$\Lambda := \{\gamma \in \Gamma \mid \gamma CU \cap CU \neq \emptyset\}$$

is finite, say of cardinality N. Then, given $\sigma \in \Gamma$, the set

$$\Lambda_\sigma = \{\gamma \in \Gamma \mid \gamma CU \cap \sigma CU \neq \emptyset\}$$

is equal to $\sigma \Lambda$, as is immediately checked, and hence also of cardinality N. Therefore, σCU meets N of the sets γCU. With (3), this implies for $x \in C$ that

$$\tag{4} \sum_{\gamma \in \Gamma} |\varphi(\gamma x)| \leqslant N\|\alpha\|_\infty \int_{\Gamma \backslash G} |\varphi(y)|\,dy < \infty$$

(since $\varphi \in L^1(\Gamma \backslash G)$), which proves the absolute and uniform convergence on compact sets.

The function p_φ is obviously Γ-invariant on the left and K-finite on the right (of the same type as φ). In view of the relation $R\varphi = \varphi * R\alpha$ ($R \in \mathcal{U}(\mathfrak{g})$) this also applies to $R\varphi$. The series can therefore be differentiated term by term up to any order, hence p_φ is also $\mathcal{Z}$-finite (of the same type as φ). Thus, it fulfills the first three conditions of 5.5. We have

(5) $$\int_G |\varphi(y)| \, dy = \int_{\Gamma \backslash G} d\bar{y} \, p_{|\varphi|}(\bar{y})$$

by the Fubini theorem; therefore, $p_\varphi \in L^1(\Gamma \backslash G)$. That it is an automorphic form now follows from 5.8.

(ii) There exists a symmetric open relatively compact neighborhood U of 1 such that $\Gamma \cap U^2 = \{1\}$. In particular,

(6) $$U\gamma \cap U\sigma = \emptyset \quad (\gamma, \sigma \in \Gamma, \, \gamma \neq \sigma).$$

By 2.14, we can find $\alpha \in I_c^\infty(G)$, with support in U, such that $\varphi = \alpha * \varphi$. The equalities

(7) $$\varphi(\gamma.x) = \int_G \alpha(\gamma xy)\varphi(y^{-1}) \, dy = \int_G \alpha(z)\varphi(z^{-1}\gamma x) \, dz \quad (x \in G)$$

and the fact that $\operatorname{supp} \alpha \subset U$ yield

(8) $$|\varphi(\gamma x)| \leqslant \|\alpha\|_\infty \int_{U\gamma x} |\varphi(y)| \, dy \quad (x \in G).$$

By (6), the $U\gamma$ are disjoint, hence so are the $U\gamma x$ for a given x. Hence (8) implies

(9) $$\sum_{\gamma \in \Gamma} |\varphi(\gamma x)| \leqslant \|\alpha\|_\infty \int_G |\varphi(y)| \, dy = \|\alpha\|_\infty \|\varphi\|_1 < \infty$$

for any $x \in G$, which proves (ii).

Remark. For a general form of this argument, due to Harish-Chandra, see [2, pp. 187–88] or [4], wherein uniform convergence on compact sets is also proved in case (ii). This implies in the latter case that p_φ is also $\mathcal{Z}$-finite. Note that p_φ is not necessarily K-finite on either side.

6.2 Theorem. *Let m be an integer $\geqslant 4$. Let f be a bounded holomorphic function on the open unit disc* D. *Then the series*

(1) $$P_f(w) = \sum_{\gamma \in \Gamma} \mu_D(\gamma, w)^{-m} f(\gamma.w)$$

converges absolutely and locally uniformly and defines a holomorphic automorphic form of weight m. The function $g \mapsto \mu_D(g, 0)^{-m} P_f(g.0)$ is in $L^1(\Gamma \backslash G)$ and is bounded if f is a polynomial.

Proof. Let

(2) $$\varphi(g) = f(g.0)\mu_D(g, 0)^{-m}.$$

By 5.15, φ is $\mathcal{Z}$-finite; in fact, φ is an eigenfunction for C with eigenvalue $(m^2/2) - m$ and K-finite on the right of type m. Since f is bounded and (by assumption) $m \geqslant 4$, $\varphi \in L^1(G)$ by 4.5. Therefore 6.1(i) implies that

$$(3) \qquad\qquad p_\varphi(g) = \sum_{\gamma \in \Gamma} \varphi(\gamma.g)$$

converges absolutely and locally uniformly and represents an automorphic form that belongs to $L^1(\Gamma \backslash G)$. But we have

$$(4) \qquad\qquad P_f(g.0) = \mu_D(g, 0)^m . p_\varphi(g),$$

which – together with the cocycle identity – implies (1). As in the case of the upper half-plane, $\mu_D(g, 0)$ has moderate growth and hence so does P_f; this implies A4. This proves 6.2 except for the last assertion. If f is a polynomial in w, then φ is left K-finite by 5.17, and we may apply 6.1(ii). This implies the last assertion of the theorem.

6.3 Let $m = 2n$ be even. In view of the relation $\frac{dg}{dw} = \mu_D(g, w)^{-2}$ (see 4.1(5) and (6)), we can write $P_f(w)$ as

$$(1) \qquad\qquad P_f(w) = \sum_{\gamma \in \Gamma} \left(\frac{d\gamma}{dw}(w) \right)^n f(\gamma.w).$$

These are the series introduced by Poincaré, but nowadays *all* the series considered in this section are called Poincaré series. We now sketch the argument of Poincaré to prove that (1) converges absolutely and uniformly on compact sets for $n \geqslant 2$.

Since f is bounded, it suffices to prove this assertion for

$$(2) \qquad\qquad Q(w) = \sum_{\gamma \in \Gamma} \left| \frac{d\gamma}{dw}(w) \right|^n .$$

In fact, we prove it slightly more generally for

$$(3) \qquad\qquad Q_m(w) = \sum_{\gamma \in \Gamma} |\mu_D(\gamma, w)|^{-m} \quad (m \geqslant 4).$$

We consider first the case $m = 4$. Fix $\delta \in (0, 1)$, and let C be the closed disc with radius δ centered at the origin. Let

$$M_{C,g} = \max_{w \in C} |\mu_D(g, w)|,$$
$$m_{C,g} = \min_{w \in C} |\mu_D(g, w)|.$$

We claim that

$$(4) \qquad\qquad M_{C,g} \leqslant 2(1 - \delta)^{-1}.m_{C,g} \quad (g \in G).$$

Write $g = \begin{pmatrix} a & b \\ \bar{b} & \bar{a} \end{pmatrix}$ as usual. If $b = 0$ (i.e., if $g \in K_D$) then $\mu_D(g, w) = \bar{a}$ is independent of $w \in D$, so

(5) $M_{c,g} = m_{C,g} \quad (g \in K_D).$

Assume now that $b \neq 0$. Then

$$M_{C,g} = \max_{w \in C} |b||w + a/b| \leqslant \max_{w \in C} |b|(|w| + |a/b|) \leqslant |b|(1 + |a/b|),$$
$$m_{C,g} = \min_{w \in C} |b||w + a/b|.$$

Because $|b||w + a/b| \geqslant (|a/b| - |w|)|b|$ (note that $|a/b| \geqslant 1$ and $|w| < 1$), we have

$$m_{C,g} \geqslant \min_{w \in C} |b|(|a/b| - |w|) = |b|(|a/b| - \delta).$$

Our assertion in the case $b \neq 0$ now follows since $|a/b| \geqslant 1$. But $2(1 - \delta)^{-1} \geqslant 1$, so the inequality (4) for $b = 0$ follows from (5). This proves (4).

In the remainder of the proof, we let $m(A)$ denote the Euclidean measure, with respect to $du \wedge dv$, of a measurable subset A of D.

For $w_0 \in C$ and $\gamma \in \Gamma$, we obviously have

$$|\mu_D(\gamma, w_0)|^{-4} = m(C)^{-1} \int_C |\mu_D(\gamma, w_0)|^{-4} \, du \, dv,$$

which, by (4), implies

(6) $|\mu_D(\gamma, w_0)|^{-4} \leqslant 2(1 - \delta)^{-1} m(C)^{-1} \int_C |\mu_D(\gamma, w)|^{-4} \, du \, dv.$

Since $|\mu_D(\gamma, w)|^{-4} = \left| \frac{d\gamma}{dw}(w) \right|^2$, the last integral is equal to $m(\gamma.C)$; hence $Q_4(w_0)$ is majorized termwise by

$$R_C = 2(1 - \delta)^{-1} m(C)^{-1} \left(\sum_{\gamma \in \Gamma} m(\gamma.C) \right)$$

for all $w_0 \in C$. Therefore it suffices to prove that the last sum converges. The set

$$\Lambda = \{ \gamma \in \Gamma \mid \gamma C \cap C \neq \emptyset \}$$

is finite and, for $\sigma \in \Gamma$,

$$\Lambda_\sigma = \{ \gamma \in \Gamma \mid \gamma C \cap \sigma C \neq \emptyset \}$$

is equal to $\sigma \Lambda$. If N is the number of elements in Λ, then

(7) $\sum m(\gamma(C)) \leqslant N m(D) < \infty.$

This proves our assertion for $m = 4$. It also shows, given C, that there exists a finite subset E of Γ such that

$$|\mu_D(\gamma, w)|^{-4} < 1 \quad \text{for all} \ \gamma \in \Gamma - E, \ w \in C.$$

Therefore

$$|\mu_D(\gamma, w)|^{-m} \leqslant |\mu_D(\gamma, w)|^{-4} \quad (\gamma \in \Gamma - E, \ w \in C),$$

so $Q_m(w)$ is majorized termwise on C by $c_m R_C$ for some positive constant c_m. The theorem follows.

6.4 Remarks. (1) In these proofs of convergence, the fact that Γ is of finite co-volume (a standing assumption here) was never used. The proofs are valid for any discrete subgroup of G.

(2) It may happen that a Poincaré series is identically zero. However, it can be shown that, for suitable weights, there exist nonzero Poincaré series. More precisely, let $a_1, \ldots, a_q \in D$ be distinct, let f_i be a function holomorphic in a neighborhood of a_i $(i = 1, \ldots, q)$, and let $d \in \mathbb{N}$. Then there exists m for which one can find q Poincaré series $P_1, \ldots, P_q$ of weight m such that the Taylor developments of P_i and f_i at a_i coincide up to order d $(1 \leqslant i \leqslant q)$.

Assume now that $P_1, \ldots, P_q$ have no common zero, and consider the map $\varphi \colon D \to \mathbb{C}^q$ defined by $\varphi(w) = (P_1(w), \ldots, P_q(w))$. In view of our assumption, $\varphi(w) \neq 0$ and so defines a point $[\varphi(w)]$ in the projective space $\mathbb{P}^{q-1}(\mathbb{C})$. If $\gamma \in \Gamma$, we have

$$\varphi(\gamma.w) = \mu_D(\gamma, w)^m \varphi(w)$$

and therefore $[\varphi(\gamma w)] = [\varphi(w)]$, so that φ induces holomorphic mapping

$$\tilde{\varphi} \colon \Gamma \backslash D \to \mathbb{P}^{q-1}(\mathbb{C}).$$

If now $\Gamma \backslash D$ is compact, this last assumption is fulfilled for a suitable m; in fact, one can show the existence of a map $\tilde{\varphi}$ that provides a projective embedding of $\Gamma \backslash D$.

For all this, see for example Chapter 5 in [2] and the references given there.

(3) The quotient of two holomorphic automorphic forms of the same weight is invariant under Γ and hence defines a meromorphic function on $\Gamma \backslash D$. Poincaré introduced the series 6.3(1) precisely to construct meromorphic functions on $\Gamma \backslash D$ in this way. Note that, in view of the regularity conditions at the cusps, they extend to meromorphic functions on the natural compactification of $\Gamma \backslash D$ (see 3.19).

7 Constant term. The fundamental estimate

7.1 Constant term. Let $P = N.A$ be a cuspidal parabolic subgroup for Γ and f a continuous Γ_N-invariant function on G that is locally L^1. The constant term f_P of f along P is, by definition, the function

$$(1) \qquad f_P(g) = \int_{\Gamma_N \backslash \Gamma} f(ng)\, dn \quad (g \in G),$$

where dn is normalized by the condition

$$(2) \qquad \int_{\Gamma_N \backslash \Gamma} dn = 1.$$

Clearly, f_P is N-invariant on the left and

$$(3) \qquad (f_P)_P = f_P.$$

The constant term is defined by integration on the left. Therefore, this operation commutes with the action of left-invariant differential operators, that is,

$$(4) \qquad D(f_P) = (Df)_P \quad (D \in \mathcal{U}(\mathfrak{g}),\ f \in C^\infty(\Gamma_N \backslash G)),$$

and with convolution on the right:

$$(5) \qquad (f * \varphi)_P = f_P * \varphi \quad (\varphi \in C_c^\infty(G)),$$

as follows easily from the definition.

7.2 Consider now the classical case (5.7), where

$$P = P_0, \quad \Gamma_N = \langle T_c \rangle, \quad f(g) = \mu(g, i)^{-m} F(g.i),$$

and $F(z)$ has the convergent power series development

$$F(z) = \sum_{n \geqslant 0} a_n e^{2\pi i n z/c} = \sum_{n \geqslant 0} a_n e^{2\pi i n x/c} . e^{-2\pi n y/c}.$$

Since y is invariant under N_0 and $dn = c^{-1} dx$, we have

$$f_P(g) = \mu(g, i)^{-m} \sum_{n \geqslant 0} a_n e^{-2\pi n y/c} \int_0^c e^{2\pi i n x/c} c^{-1}\, dx.$$

In view of

$$c^{-1} \int_0^c e^{2\pi i n x/c}\, dx = \begin{cases} 0 & n \neq 0, \\ 1 & n = 0, \end{cases}$$

the constant term is

$$(1) \qquad f_P(g) = \mu(g, i)^{-m} a_0$$

70

and the constant term of F, which is by definition

(2) $$F_P(z) = c^{-1} \int_0^c F(z+x)\,dx = a_0,$$

is really a constant, which explains the terminology. Moreover, we can write

(3) $$|f_P(g)| = |a_0| y^{m/2}$$

in view of (3) and (6) in 3.3.

7.3 We return to the general case of 7.1 and let Y be an element of the Lie algebra of N such that e^Y generates the (infinite cyclic) group Γ_N. For $a \in A$, we have (cf. 2.8)

$$\text{Ad}\,a(Y) = a^{2\rho}.Y.$$

This means, in the usual terminology, that the character 2ρ of A is the simple root of $\mathfrak{g}$ with respect to the Lie algebra $\mathfrak{a}$ of A, for the ordering defined by P, and we shall accordingly denote it by α.

7.4 Main lemma. *Let X_1, X_2, X_3 be a basis of the Lie algebra $\mathfrak{g}$ of G. Let $f \in C^1(\Gamma_N \backslash G)$. Then there exists $c > 0$, independent of f, such that*

(1) $$|(f - f_P)(x)| \leqslant c.a(x)^{-\alpha} \left(\sum_1^3 |X_i f|_P(x) \right) \quad (x \in G).$$

Note that X_i is identified with a left-invariant differential operator, so that $X_i f$ is still left-invariant under Γ_N and the constant term of $|X_i f|$ is well-defined.

Proof. In view of 7.1(1) and (2), we have

(2) $$(f_P - f)(x) = \int_0^1 \left(f(e^{tY}x) - f(x) \right) dt.$$

But, clearly,

(3) $$f(e^{tY}x) - f(x) = \int_0^t \frac{d}{ds} f(e^{(u+s)Y}.x) \bigg|_{s=0} du.$$

In the notation of 2.2, this can be written as

(4) $$f(e^{tY}x) - f(x) = \int_0^t ((-Y) * f)(e^{uY}x)\,du;$$

that is, we have used the *right*-invariant differential operator on G defined by $-Y$ (see 2.2). We want to express (4) in terms of left-invariant differential operators. We have

$$\frac{d}{dt} f(e^{tY}x)\Big|_{t=0} = \frac{d}{dt} f(x.x^{-1}.e^{tY}x)\Big|_{t=0} = \frac{d}{dt} f(x.e^{t\operatorname{Ad}x^{-1}(Y)})\Big|_{t=0}.$$

By the Iwasawa decomposition (2.7),

$$x = n_x.a(x)k_x \quad (n_x \in N,\ a(x) \in A,\ k_x \in K).$$

We have

$$\operatorname{Ad} n_x(Y) = Y, \qquad \operatorname{Ad} a(x)^{-1}(Y) = a(x)^{-\alpha}Y$$

(cf. 2.8 for the second equality). Furthermore, there exist smooth functions c_i ($i = 1, 2, 3$) on K such that

$$\operatorname{Ad} k_x^{-1}(Y) = \sum c_i(k_x^{-1})X_i;$$

therefore,

$$(-Y * f)(x) = a(x)^{-\alpha} \sum_1^3 c_i(k_x^{-1}).X_i f(x).$$

The functions $|c_i|$ are bounded on K, since the latter is compact. Let then

$$c \geqslant \max_{i,k} |c_i(k)| \quad (i = 1, 2, 3, k \in K).$$

Then

$$|(-Y * f)(e^{sY}x)| \leqslant ca(x)^{-\alpha} \sum |X_i f(e^{sY}.x)|$$

since, obviously, $a(x) = a(e^{sY}x)$ for all $x \in G$ and $s \in \mathbb{R}$. The inequality

$$|f(e^{tY}x) - f(x)| \leqslant \int_0^1 |(-Y * f)(e^{sY}x)|\, ds,$$

which follows from (4), then yields

$$|f(e^{tY}x) - f(x)| \leqslant ca(x)^{-\alpha} \sum \int_0^1 |X_i f|(e^{sY}x)\, ds = ca(x)^{-\alpha} \sum |X_i f|_P(x),$$

and the lemma now follows from (2).

For later use, we draw a first consequence of the main estimate in a form more general than needed for the immediate applications to automorphic forms.

7.5 Theorem. *Let P be a Γ-cuspidal parabolic subgroup. Let ψ be a smooth function on G, left-invariant under Γ_N, of uniform moderate growth (see 5.1) on some Siegel set $\mathfrak{S}_t$ with respect to P. Then $\psi - \psi_P$ is rapidly decreasing on $\mathfrak{S}_t$.*

Proof. By assumption, there exists $\lambda \in X(A)_{\mathbb{R}}$ such that

$$(1) \qquad |D\psi(x)| \prec a(x)^{\lambda} \quad (x \in \mathfrak{S}_t,\ D \in \mathcal{U}(\mathfrak{g})).$$

Applying this to $D = X_1, X_2, X_3$ we have, from 7.4,

(2) $$|(\psi - \psi_P)(x)| \prec a(x)^{\lambda - \alpha} \quad (x \in \mathfrak{S}_t).$$

For any locally L^1 function η that is left Γ_N-invariant, we have $(\eta - \eta_P)_P = 0$ (see 7.1(3)). Applying the lemma repeatedly to $\psi - \psi_P$ we obtain

(3) $$|(\psi - \psi_P)(x)| \prec a(x)^{\lambda - m\alpha} \quad (x \in \mathfrak{S}_t)$$

for *any* $m \in \mathbb{N}$. Given $\mu \in X(A)_{\mathbb{R}}$, a real valued character of A (see 2.8), there exists $m \in \mathbb{N}$ such that

$$\mu = \lambda - m\alpha + r.\rho \quad \text{with } r > 0.$$

Then

$$a(x)^{\lambda - m\alpha} = a(x)^{\mu - r\rho} \prec a(x)^{\mu};$$

hence

(4) $$|(\psi - \psi_P)(x)| \prec a(x)^{\mu} \quad (x \in \mathfrak{S}_t)$$

for any μ, which means that $\psi - \psi_P$ is rapidly decreasing on $\mathfrak{S}_t$.

7.6 Corollary. (i) *Let f be a locally L^1 function on G that is left-invariant under Γ_N, of moderate growth on $\mathfrak{S}_t$, and let $\varphi \in C_c^\infty(G)$. Then $f * \varphi - (f * \varphi)_P$ is rapidly decreasing on $\mathfrak{S}_t$.*

(ii) *Let f be an automorphic form for Γ, and let P be a Γ-cuspidal parabolic subgroup. Then $f - f_P$ is rapidly decreasing on a Siegel set with respect to P.*

Proof. (i) The convolution $f * \varphi$ is left-invariant with respect to Γ_N, and is smooth and of uniform moderate growth on $\mathfrak{S}_t$ (5.3); it therefore satisfies all the conditions imposed on ψ in 7.5.

(ii) By assumption, f is of moderate growth on $\mathfrak{S}_t$ (5.5). Moreover, there exists $\varphi \in C_c^\infty(G)$ such that $f = f * \varphi$ (5.6(b)). Therefore (ii) follows from (i).

As in 3.18, let P_i $(i = 1, \ldots, l)$ be a set of representatives of Γ-conjugacy classes of cuspidal parabolic subgroups. Let $\mathfrak{S}_i$ be a Siegel set with respect to P_i, so that the union of the $\mathfrak{S}_i$ with a compact set is a fundamental set Ω for Γ in G.

7.7 Corollary. *Let $p \in [1, \infty]$ and let f be an automorphic form. Then the following conditions are equivalent*:

(i) $f \in L^p(\Gamma \backslash G)$;

(ii) *for any Siegel set $\mathfrak{S}$ with respect to a cuspidal parabolic subgroup,*

$$f_P \in L^p(\mathfrak{S});$$

(iii) $f_{P_i} \in L^p(\mathfrak{S}_i)$ $(i = 1, \ldots, l)$.

That (i) $\Rightarrow$ (ii) $\Rightarrow$ (iii) is clear. Assume (iii). By 7.6, $f - f_{P_i}$ is fast decreasing on $\mathfrak{S}_i$, so $f | \mathfrak{S}_i$ belongs to $L^p(\mathfrak{S}_i)$. Since the $\mathfrak{S}_i$ and a compact set cover $\Gamma \backslash G$, condition (i) follows.

7.8 Definition. Let f be a locally L^1 function on $\Gamma \backslash G$. It is said to be *cuspidal* for Γ if its constant term with respect to every cuspidal parabolic subgroup for Γ is zero. If V is a space of locally L^1 functions on $\Gamma \backslash G$, we let $^\circ V$ denote the subspace of cuspidal elements in V.

Let $\sigma \in \Gamma$ and $P' = {}^\sigma P$. Since obviously, $^\sigma N = N'$, the automorphism Int σ induces an isomorphism of $\Gamma_N \backslash N$ onto $\Gamma_{N'} \backslash N'$, which transforms dn onto dn' (in view of 7.1(2)). We have

(1) $f_P(x) = f_{P'}(\sigma x) \quad (x \in G)$,

as follows from

$$\int_{\Gamma_N \backslash N} f(nx)\, dn = \int_{\Gamma_N \backslash N} f(\sigma n \sigma^{-1} \sigma x)\, dn = \int_{\Gamma_{N'} \backslash N'} f(n' \sigma x)\, dn'.$$

As a consequence, to show that f is cuspidal it suffices to check the vanishing of its constant term along a set of representatives of Γ-conjugacy classes of cuspidal parabolic subgroups. Corollary 7.6 clearly implies the following corollary.

7.9 Corollary. *A cuspidal automorphic form for Γ is rapidly decreasing on any Siegel set with respect to a cuspidal parabolic subgroup.*

7.10 Corollary. *Let f be a holomorphic automorphic form of weight $m \geqslant 1$ (see 5.14). Then f is cuspidal if and only if $f \in L^2(\Gamma \backslash G)$.*

If f is cuspidal, then it is fast decreasing (7.9) and thus in $L^2(\Gamma \backslash G)$. Assume that $f \in L^2(\Gamma \backslash G)$. Then its constant term f_P with respect to a cuspidal pair (P, A) is also square integrable on a Siegel set (7.7). After conjugation, we may assume that ∞ is a cuspidal point and that $(P, A) = (P_0, A_0)$. Then, in the notation of 7.2(3), $|f_P(g)| = |a_0|.y^{m/2}$. We take as a Siegel set a rectangle $|x| \leqslant a$, $y > b$; the invariant measure is $y^{-2}.dx\,dy$. Because $m - 2 \geqslant -1$, $|f_P|^2$ is not integrable unless $a_0 = 0$; hence f is cuspidal.

8 Finite dimensionality of the space of automorphic forms of a given type

8.1 By assumption, the quotient $\Gamma\backslash G$ has finite volume, say C. This implies that, for $p \geq 1$, the space $L^p(\Gamma\backslash G)$ is contained in $L^1(\Gamma\backslash G)$ and the injection is continuous. In fact, by the Hölder inequality we have

$$(1) \qquad \int_{\Gamma\backslash G} |f(x)|\,dx \leq \left(\int_{\Gamma\backslash G} |f(x)|^p\,dx\right)^{1/p} \cdot \left(\int_{\Gamma\backslash G} dx\right)^{1/q},$$

where $q^{-1} = 1 - p^{-1}$; that is,

$$(2) \qquad \|f\|_1 \leq C^{1/q} \cdot \|f\|_p \quad (f \in L^p(\Gamma\backslash G)).$$

We let $^{\circ}L^p(\Gamma\backslash G)$ be the space of cuspidal elements in $L^p(\Gamma\backslash G)$. Clearly

$$(3) \qquad {}^{\circ}L^p(\Gamma\backslash G) = L^p(\Gamma\backslash G) \cap {}^{\circ}L^1(\Gamma\backslash G) \quad (p \geq 1).$$

8.2 Proposition. *The subspace $^{\circ}L^p(\Gamma\backslash G)$ is closed in $L^p(\Gamma\backslash G)$ $(p \geq 1)$.*

In view of 8.1(3), it suffices to prove this assertion for $p = 1$.

Given a Γ-cuspidal parabolic subgroup P and $\varphi \in C_c^{\infty}(N\backslash G)$, we let

$$\lambda_{P,\varphi}(f) = \int_{\Gamma_N\backslash G} f(x)\varphi(x)\,dx.$$

We then have

$$\lambda_{P,\varphi}(f) = \int_{N\backslash G} \varphi(\bar{x})\,d\bar{x} \int_{\Gamma_N\backslash N} f(nx)\,dn = \int_{N\backslash G} \varphi(\bar{x})f_P(\bar{x})\,d\bar{x},$$

where $\bar{x}$ is the projection in $N\backslash G$ of $x \in \Gamma_N\backslash G$ and $d\bar{x}$ the quotient measure on $N\backslash G$ (see 2.9). Therefore $f_P = 0$ if and only if $\lambda_{P,\varphi}(f) = 0$ for all $\varphi \in C_c^{\infty}(N\backslash G)$. As a consequence, $^{\circ}L^1(\Gamma\backslash G)$ is the intersection of the kernels of the linear forms $\lambda_{P,\varphi}$, as P runs through the cuspidal parabolic subgroups and φ through $C_c^{\infty}(N_P\backslash G)$. It therefore suffices to show that $\lambda_{P,\varphi}$ is a continuous linear form on $L^1(\Gamma\backslash G)$.

The support of φ in $N\backslash G$ is compact. Its inverse image in G is the support of φ, viewed as a function on G, which can be written as $N.D$ with D compact; since Γ_N is cocompact in N, it can also be written as $\Gamma_N.E$ with E compact. Then

$$|\lambda_{P,\varphi}(f)| \leq \int_E |f(x)||\varphi(x)|\,dx \leq \|\varphi\|_{\infty} \int_E |f(x)|\,dx.$$

There exist finitely many compact neighborhoods $U_1, \ldots, U_m$ of points in E such that E is contained in the union of the U_i and $G \to \Gamma\backslash G$ is a homeomorphism of U_i onto its image for all i. Therefore

$$\int_E |f(x)|\, dx \leqslant m \int_{\Gamma \backslash G} |f(x)|\, dx;$$

hence

$$|\lambda_{P,\varphi}(f)| \leqslant \|\varphi\|_\infty m \|f\|_1,$$

which shows that $\lambda_{P,\varphi}$ is continuous on $L^1(\Gamma \backslash G)$.

8.3 Lemma. *Let Z be a locally compact space with a positive measure μ such that $\mu(Z)$ is finite. Let V be a closed subspace of $L^2(Z, \mu)$ that consists of essentially bounded functions. Then V is finite dimensional.*

This lemma is due to R. Godement and the proof that follows to L. Hörmander (see [31, pp. 17, 18]).

For every $f \in L^2(Z, \mu)$ we have

(1) $$\|f\|_2 \leqslant \mu(Z) \|f\|_\infty.$$

The map $(V, \| \cdot \|_\infty) \to (V, \| \cdot \|_2)$ induced by the identity of V is therefore continuous and, obviously, bijective. As a consequence of the open mapping theorem, its inverse is also continuous (see [61, §II.5, p. 77, Cor. to Thm.]). Hence there exists $c > 0$ such that

(2) $$\|f\|_\infty \leqslant c \|f\|_2 \quad (f \in V).$$

Let $v_1, \ldots, v_n$ be an orthonormal subset of V, and let $a_i \in \mathbb{C}$ $(i = 1, \ldots, n)$. By (2) we have

(3) $$\left| \sum a_i v_i(z) \right| \leqslant c \left\| \sum a_i v_i \right\|_2 = c \left(\sum |a_i|^2 \right)^{1/2}$$

for almost all z, that is, for all z outside a set of μ-measure zero. Put $a_i = \overline{v_i(z)}$. We have

$$\sum |v_i(z)|^2 \leqslant c \left(\sum |v_i(z)|^2 \right)^{1/2}$$

and hence

$$\sum |v_i(z)|^2 \leqslant c^2.$$

Integration over Z yields

$$n \leqslant c^2 \mu(Z),$$

so that $\dim V \leqslant c^2 \mu(Z)$.

8.4 Lemma. *Let $f \in L^2(\Gamma \backslash G)$. Then f is of type $\chi \in \hat{K}$ on the right if and only if*

(1) $$\int_K f(xk) \overline{\chi(k)}\, dk = f(x) \quad (x \in \Gamma \backslash G).$$

If f is of type χ, then $f(xk) = f(x).\chi(k)$ and so $f(xk)\overline{\chi(k)} = f(x)$ is independent of k. Conversely, assume (1). For $l \in K$ and replacing the variable k by $k' = lk$, we obtain

$$\int_K f(xlk)\overline{\chi(k)}\,dk = \int_K f(xk')\chi(l)\overline{\chi(k')}\,dk'$$

$$= \chi(l) \int_K f(xk')\overline{\chi(k')}\,dk' = \chi(l)f(x).$$

8.5 Theorem. *The space* $\mathcal{A}(\Gamma, J, \chi)$ *of automorphic forms for* Γ *annihilated by a nonzero ideal* J *of* $\mathcal{Z}(\mathfrak{g})$ *and of type* $\chi \in \hat{K}$ *on the right with respect to* K *(see 5.5) is finite dimensional.*

Proof. We let $\mathcal{A}$ stand for $\mathcal{A}(\Gamma, J, \chi)$.

Let $P_1, \ldots, P_l$ be a set of representatives of the Γ-conjugacy classes of cuspidal parabolic subgroups. Let ψ be the map

$$\psi: f \mapsto (f_{P_1}, \ldots, f_{P_l}) \quad (f \in \mathcal{A}).$$

We must prove that ker ψ and Im ψ are finite dimensional.

The kernel is the space $°\mathcal{A}$ of cusp forms contained in $\mathcal{A}$ (see 7.7). Any cusp form is rapidly decreasing at all cusps (7.8), is therefore bounded on $\Gamma \backslash G$, and in particular belongs to $L^2(\Gamma \backslash G)$. In order to apply 8.3, we need to know that $°\mathcal{A}$ is closed in $L^2(\Gamma \backslash G)$, in other words, that the three conditions characterizing an element in $°\mathcal{A}$ go over to L^2 limits. Let then $f_n \to f$ ($n \in \mathbb{N}$) in $L^2(\Gamma \backslash G)$, where $f_n \in °\mathcal{A}$ for all $n \in \mathbb{N}$. This implies convergence in $L^1(\Gamma \backslash G)$ (see 8.1(2)) and hence convergence in the distribution sense. Therefore, if $Z \in \mathcal{Z}(\mathfrak{g})$ then $Zf_n \to Zf$ in L^2. This shows that f is of type J, as a distribution at first.

By 8.4,

$$\int f_n(xk)\overline{\chi(k)}\,dk = f_n(x),$$

and this condition then goes over to f. Hence f is of type χ, too.

Strictly speaking, 8.4 is an "almost everywhere statement", but now f is $\mathcal{Z}$-finite and right K-finite as a distribution (say), and this implies that f is analytic (2.13) so that $f \in \mathcal{A}$. Since $°L^2(\Gamma \backslash G)$ is closed in $L^2(\Gamma \backslash G)$ by 8.2, this shows that $°\mathcal{A}$ is closed in $L^2(\Gamma \backslash G)$. Thus all conditions of 8.3 are fulfilled and so ker ψ is finite dimensional.

Note that if $\Gamma \backslash G$ is compact then ker $\psi = \mathcal{A}$, so that the theorem is proved in that case. In general, the finite dimensionality of the image of ψ is a consequence of the following lemma.

8.6 Lemma. *Let* P *be a cuspidal parabolic subgroup for* Γ. *Then* $\{ f_P \mid f \in \mathcal{A} \}$ *is a finite dimensional vector space of functions.*

Let p be the fixed point of P on $\partial \bar{X}$, and let $g \in K$ be such that $g.\infty = p$. Then, replacing Γ by ${}^g\Gamma$, we may assume that $p = \infty$ and that P is the group of upper triangular matrices. Hence P is the standard parabolic subgroup and we use the notation E, F, H of 2.1 and 2.5. The function f_P is N-invariant on the left and of K-type χ on the right. In view of the Iwasawa decomposition $G = NAK$ (2.7), f_P is completely determined by its restriction to A. It suffices then to show that these restrictions form a finite dimensional space. The function f_P is N-invariant on the left and hence is annihilated by the *right*-invariant differential operator E_r defined by E. We claim that

(1) $$E f_P(a) = 0 \quad (a \in A),$$

where E is now, as usual, the *left*-invariant differential operator defined by E (see 2.1). We have

(2) $$f_P(ae^{tE}) = f_P(ae^{tE}a^{-1}a) = f_P(e^{t\,\mathrm{Ad}\,a(E)}a) = f_P(e^{ta^\alpha E}a),$$

where $\alpha = 2\rho$ (see 2.8). Hence

(3) $$\frac{d}{dt}f_P(ae^{tE})\Big|_{t=0} = a^\alpha \frac{d}{dt}f_P(e^{tE}.a)\Big|_{t=0} = a^\alpha \frac{d}{dt}f_P(a)\Big|_{t=0} = 0.$$

The Casimir operator $\mathcal{C}$ can be written as

$$\mathcal{C} = \tfrac{1}{2}H^2 - H + 2EF$$

(see 2.5). In view of the definition of EFf_P (see 2.1(6)), the foregoing shows that $EFf_P = 0$ on A and so

$$P(\mathcal{C})f_P = Q(H)f_P \quad \text{on } A, \quad \text{where } Q(H) = P(\tfrac{1}{2}H^2 - H).$$

We can identify $\mathbb{R}$ to A by the exponential mapping that maps t onto the diagonal matrix having diagonal entries e^t and e^{-t}. Then H is just d/dt. Therefore f_P, viewed as a function on $\mathbb{R}$ in this way, is a solution of the ordinary differential equation $Q(\frac{d}{dt})f_P = 0$ and so belongs to a finite dimensional space of functions.

8.7 We have assumed J to be arbitrary, generated by some monic polynomial P. We want to show that there would be no essential loss in generality when assuming P to be a power of $\mathcal{C} - \lambda$ ($\lambda \in \mathbb{C}$).

Let T be an indeterminate. Let $\lambda_1, \ldots, \lambda_q$ be the distinct roots of $P(T)$ and $n_1, \ldots, n_q$ their multiplicities. Let J_i be the ideal of $\mathcal{Z}(\mathfrak{g})$ generated by $(\mathcal{C} - \lambda_i)^{n_i}$ ($i = 1, \ldots, q$). Then

(1) $$A(\Gamma, J, m) = \bigoplus_i A(\Gamma, J_i, m).$$

Proof. Clearly, the left-hand side contains the right-hand side. Let

$$Q_i(\mathcal{C}) = \prod_{j \neq i}(\mathcal{C} - \lambda_j)^{n_j}.$$

The polynomials Q_i are relatively prime; therefore, there exist polynomials R_i $(i = 1, \ldots, q)$ such that $\sum_i R_i \cdot Q_i = 1$. Then

$$f = \sum_i f_i, \quad \text{where} \quad f_i = R_i(C) \cdot Q_i(C)f.$$

Obviously, $(C - \lambda_i)^{n_i} f_i = 0$, hence $f_i \in A(\Gamma, J_i, m)$. Thus f belongs to the right-hand side of (1). If

$$\sum_i f_i = 0 \quad (f_i \in A(\Gamma, J_i, m)),$$

then f_i is annihilated by the relatively prime polynomials $Q_i(C)$ and $(C - \lambda_i)^{n_i}$; hence f_i is zero. The sum is therefore direct.

8.8 Lemma. *Let $f \in C(G)$ be K-finite on the right, Z-finite, and in $L^1(G)$. Let (P, A) be a p-pair and let $N = N_P$. Then*

$$(1) \qquad \int_N f(nx)\, dn = 0 \quad (x \in G).$$

Proof. By 2.14 and 2.22, there exists $\alpha \in I_c^\infty(G)$ such that $f = f * \alpha$ and f is bounded on G. Let $d\dot{x}$ be the quotient measure on $N \backslash G$ (see 2.9). Then

$$(2) \qquad \int_G |f(x)|\, dx = \int_{N \backslash G} d\dot{x} \int_N |f(nx)|\, dn,$$

so $\int_N f(nx)\, dn$ exists. Let

$$(3) \qquad \varphi(x) = \int_N f(nx)\, dx \quad (x \in G);$$

$\varphi(x)$ is left N-invariant and right K-finite. It therefore suffices to show that its restriction to A is zero. If Q is a polynomial in C, then $Q(C)f = f * Q(C)\alpha$ also satisfies our assumptions. The argument at the beginning of 8.7 then shows that f is a direct sum of functions f_i satisfying our assumptions, where moreover f_i is annihilated by $(C - \lambda)^m$ for some $\lambda \in \mathbb{C}$ and $m \in \mathbb{N}$. This reduces us to that case. We now revert to the notation and conventions used at the end of 8.6. We identify $\mathbb{R}$ with A by the exponential map and view φ as a function on $\mathbb{R}$ satisfying the ordinary differential equation

$$(4) \qquad \left(\frac{1}{2} \frac{d^2 t}{dt^2} - \frac{d}{dt} - \lambda \right)^m \varphi = 0.$$

The function φ is bounded and, moreover, is integrable on $N \backslash G = A \times K$ with respect to the quotient measure $a^{-2}\, da\, dk$ (see 2.9). In the present notation, this means that φ is integrable on $\mathbb{R}$ with respect to the measure $e^{-2t}\, dt$. The following exercise in calculus shows that this forces φ to be zero.

Let s be a square root of $2\lambda + 1$. Then $1 + s$ and $1 - s$ are the two roots of the polynomial $T^2/2 - T - \lambda$, distinct if $s \neq 0$. By the standard theorem on the solutions of (4), (see e.g. [14, IV, no. 8]),

$$(5) \qquad \varphi(t) = p_1(t)e^{t(1+s)} + p_2(t)e^{t(1-s)} \quad (s \neq 0),$$

where $p_i(t)$ $(i = 1, 2)$ is a polynomial with constant coefficients of degree $\leqslant m$ and

$$(6) \qquad \varphi(t) = q(t).e^t$$

if $s = 0$, where $q(t)$ is a polynomial of degree $\leqslant 2m$. It is then left to the reader to check that: (a) φ bounded implies $\mathcal{R}s = -1$ and p_i constant if $s \neq 0$ and $\varphi = 0$ if $s = 0$; (b) if φ in (5) is not zero then it is not integrable with respect to $e^{-2t} dt$.

8.9 Theorem. *Let f be a function on G that is $\mathcal{Z}$-finite, K-finite on both sides, and belongs to $L^1(G)$. Then the Poincaré series p_f (see 6.1) is a cusp form.*

Proof. Let (P, A) be a cuspidal p-pair. Then

$$(p_f)_P(x) = \int_{\Gamma_N \backslash N} p_f(nx)\, dn = \int_{\Gamma_N \backslash N} dn \left(\sum_\gamma f(\gamma nx) \right)$$

$$= \int_N dn \sum_{\Gamma_N \backslash \Gamma} f(\gamma nx) = \sum_{\Gamma_N \backslash N} \int_N f(\gamma nx)\, dn.$$

For any $y \in G$, the function $x \mapsto f(y.x)$ satisfies the assumptions of 8.8; hence the latter implies that each term on the right-hand side is zero.

Remarks. (1) In the classical case of a holomorphic Poincaré series, 8.9 already follows from 7.2(3) and the fact that p_φ is bounded on $\Gamma \backslash G$.

(2) The holomorphic Poincaré series have been introduced in the unit disc model. It is also possible to define them directly on the upper half-plane and show that they span the space of holomorphic cusp forms (see [36, 9.5] or [48, 1.5]).

9 Convolution operators on cuspidal functions

9.1 Compact operators [61, X.5; 46, VI.5]. Let H be a Hilbert space (with a countable basis), $(\ ,\ )$ the scalar product on H, and $\mathcal{L}(H)$ the algebra of bounded linear operators on H. If $A \in \mathcal{L}(H)$ then A^* denotes its adjoint – that is, the unique bounded linear operator such that $(Ax, y) = (x, A^*y)$ $(x, y \in H)$. The operator A^*A is self-adjoint and positive (i.e., $(A^*Ax, x) \geqslant 0$ for all $x \in H$); it has a unique positive square root, called the *absolute value* $|A|$ of A.

The bounded operator A is *compact* if it transforms any bounded set into a relatively compact one. The eigenspaces of A corresponding to nonzero eigenvalues are finite dimensional. If A is compact, positive, and self-adjoint, then H has an orthonormal basis $\{e_i\}$ consisting of eigenvectors of A: $Ae_i = \lambda_i e_i$ with $\lambda_i \to 0$. The operator A is compact if and only if $|A|$ is. The compact operators obviously form an ideal in $\mathcal{L}(H)$.

More generally, a continuous linear map $A: H \to H'$ of H into a Hilbert space H' is said to be compact if it transforms any bounded set into a relatively compact one or, equivalently, any bounded sequence into one containing a convergent subsequence.

The notion of compact operator can be defined similarly in any topological vector space with a topology defined by seminorms, and in particular in an L^p-space: such an operator transforms any set bounded with respect to the L^p-norm into a relatively compact one.

9.2 Theorem. *Fix $p \geqslant 1$ $(p \in \mathbb{R})$ and $\varphi \in C_c^1(G)$. There exists a constant $C(\varphi) > 0$ such that*

$$(1) \qquad |(f * \varphi)(x)| \leqslant C(\varphi)\|f\|_p \quad (x \in \Gamma \backslash G)$$

for all $f \in {}^\circ L^p(\Gamma \backslash G)$.

Proof. Let P be a cuspidal parabolic subgroup for Γ and $\mathfrak{S}$ a Siegel set relative to P. By 5.7, there exists a constant c_1, depending only on φ, such that

$$(2) \qquad |(f * \varphi)(x)| \leqslant c_1 \|f\|_p a(x)^{2\rho/p} \quad (x \in \mathfrak{S},\ f \in L^p(\Gamma \backslash G)).$$

For $D \in \mathcal{U}(\mathfrak{g})$ we have $D(f * \varphi) = f * D\varphi$; hence by 5.7 there also exists a constant $c(\varphi, D)$ such that

$$(3) \qquad |(D(f * \varphi))(x)| \leqslant c(\varphi, D)\|f\|_p a(x)^{2\rho/p} \quad (x \in \mathfrak{S})$$

for all $f \in L^p(\Gamma \backslash G)$.

From the definition of the constant term, and since $a(nx) = a(x)$ for $n \in N$ and $x \in G$, we also have

(4) $|D(f * \varphi)_P(x)| \leqslant c(\varphi, D) \|f\|_p a(x)^{2\rho/p}$ $(x \in \mathfrak{S}, f \in L^p(\Gamma \backslash G))$.

Assume now that f is moreover cuspidal at P, that is, $f_P = 0$. Then $(f * \varphi)_P = 0$ (see 7.1(5)). It then follows from 7.4 and (4) applied to

$$D = X_1, X_2, X_3$$

that there exists a constant $c_2(\varphi)$, depending on φ but not on f, such that

(5) $|(f * \varphi)(x)| \leqslant c_2(\varphi) a(x)^{-\alpha} \|f\|_p a(x)^{2\rho/p}$ $(x \in \mathfrak{S}, f \in L^p(\Gamma \backslash G))$.

As in 7.5, we can iterate the application of 7.4. It follows that, given $m \in \mathbb{N}$, there exists a constant $c(m, \varphi)$ such that

(6) $|(f * \varphi)(x)| \leqslant c(m, \varphi) \|f\|_p a(x)^{2\rho/p} . a(x)^{-m\alpha}$ $(x \in \mathfrak{S}, f \in L^p(\Gamma \backslash G))$.

Recall that $\alpha = 2\rho$. Therefore, if $m \geqslant 2$ then

(7) $a(x)^{2\rho/p - m\alpha} \leqslant 1$ $(x \in \mathfrak{S})$

and (1) follows from (6) and (7) for $x \in \mathfrak{S}$. If now $f \in {}^\circ L^p(\Gamma \backslash G)$, then this estimate is valid on a Siegel set relative to any cuspidal parabolic subgroup for Γ. Since $\Gamma \backslash G$ is covered by finitely many such sets, (1) is established.

9.3 Theorem.   *Let $p \in \mathbb{R}$, $p \geqslant 1$, and $\varphi \in C_c^1(G)$. Then the operator $*\varphi$ on ${}^\circ L^p(\Gamma \backslash G)$ is compact.*

Proof. Let $D \in \mathcal{U}(\mathfrak{g})$. Since $D(f * \varphi) = f * D\varphi$, 9.2 also implies the existence of a constant $c(\varphi, D)$ such that

(1) $|D(f * \varphi)(x)| \leqslant c(\varphi, D) \|f\|_p$ $(x \in \Gamma \backslash G, D \in \mathcal{U}(\mathfrak{g}), f \in {}^\circ L^p(\Gamma \backslash G))$.

Fix an open subset U of $\Gamma \backslash G$ that is the homeomorphic image of a (sufficiently small) coordinate neighborhood in G, with coordinates x_1, x_2, x_3. The partial derivatives of any order with respect to the x_i are, in U, linear combinations of elements in $\mathcal{U}(\mathfrak{g})$ with smooth coefficients (see 2.1); therefore, (1) is also valid for the partial derivatives of any order if $x \in U$.

Let now $\{f_n\}$ $(n = 1, 2, \ldots)$ be a sequence of elements in ${}^\circ L^p(\Gamma \backslash G)$, with $\|f_n\|_p \leqslant a$, for some constant $a > 0$. Then the previous remark shows that the derivatives of any order of the $f_n * \varphi$ are uniformly bounded in U. Therefore the $f_n * \varphi$, or their derivatives of any order, form an equicontinuous family. By Ascoli's theorem, this family has a uniformly convergent subsequence. Using the diagonal process over a countable cover of $\Gamma \backslash G$ by sets such as U, we can there-fore extract from the f_n a subsequence $f_{n'}$ such that $f_{n'} * \varphi$ converges locally uni-formly to a function f, which is then continuous and bounded by a. Since $\Gamma \backslash G$ has finite measure, it follows readily that $f \in L^p(\Gamma \backslash G)$ and $f_{n'} \to f$ in $L^p(\Gamma \backslash G)$. By 8.2, $f \in {}^\circ L^p(\Gamma \backslash G)$, and the theorem is proved.

Remark. For $p = 2$ (the only case of interest in the sequel), this theorem, in the general case, is due to Gelfand and Piatetski-Shapiro (see [21]). Compactness (in a bigger space, however; see 9.7, but only for $\varphi \in C_c^\infty(G)$) is all that we need, but in fact more is true – namely, that $*\varphi$ is of trace class. For the sake of com-pleteness, we sketch a proof of this result after first recalling some definitions and facts about two types of bounded operators.

9.4 Hilbert–Schmidt and trace class operators. Let H, $(\,,\,)$, and $\mathcal{L}(H)$ be as in 9.1. The trace $\mathrm{tr}_\mathcal{E} A$ of a bounded operator A with respect to an orthonormal basis $\mathcal{E} = \{e_i\}$ is the sum of the (Ae_i, e_i). A positive operator is of *trace class* if that sum is finite. It is then so for every orthonormal basis; the trace is indepen-dent of the basis and is denoted $\mathrm{tr}\, A$. A bounded operator A is of trace class if $|A|$ is. It is then compact, and this condition amounts to the convergence of the series $\sum \lambda_i$ (with the λ_i as in 9.1).

A bounded operator is *Hilbert–Schmidt* if A^*A is of trace class. This implies compactness, and is equivalent to the convergence of $\sum \lambda_i^2$.

Let $\mathcal{L}_1(H)$ (resp. $\mathcal{L}_2(H)$) be the set of trace class (resp. Hilbert–Schmidt) op-erators on H. These are ideals in $\mathcal{L}(H)$ and $\mathcal{L}_1(H) \subset \mathcal{L}_2(H)$. Moreover, A is of trace class if and only if it is a product of two Hilbert–Schmidt operators.

For all this, see Chapter VI in [46].

9.5 Theorem. *Let $\varphi \in C_c^\infty(G)$. Then $*\varphi$ is of trace class on ${}^\circ L^2(\Gamma \backslash G)$.*

Proof. By 9.2(1) there exists a constant $C > 0$ such that

$$|(f * \varphi)(x)| \leqslant C \|f\|_2 \quad (x \in \Gamma \backslash G, \; f \in {}^\circ L^2(\Gamma \backslash G)).$$

For a given $x \in G$, this shows that $f \mapsto f * \varphi(x)$ is a continuous linear form on ${}^\circ L^2(\Gamma \backslash G)$. Hence there exists an element $k_x \in {}^\circ L^2(\Gamma \backslash G)$ such that

$$f * \varphi(x) = (f, k_x) = \int_{\Gamma \backslash G} f(y)\overline{k_x(y)}\, dy.$$

In particular,

$$\|k_x\|_2^2 = (k_x, k_x) = (k_x * \varphi)(x) \leqslant |(k_x * \varphi)(x)| \leqslant C \|k_x\|_2,$$

so it follows that

$$\|k_x\|_2 \leqslant C.$$

Write $k(x, y) = k_x(y)$ $(x, y \in \Gamma\backslash G)$. Then

$$\int_{\Gamma\backslash G}\int_{\Gamma\backslash G} |k(x, y)|^2\, dx\, dy = \int_{\Gamma\backslash G} dx \int_{\Gamma\backslash G} |k_x(y)|^2\, dy \leqslant C^2 \int_{\Gamma\backslash G} dx$$

is finite and

$$(f * \varphi)(x) = \int_{\Gamma\backslash G} f(y)\overline{k(x, y)}\, dy.$$

Therefore, $*\varphi$ on $°L^2(\Gamma\backslash G)$ is represented by a L^2-kernel. This implies that $*\varphi$ is Hilbert–Schmidt [46, VI.23].

By a theorem of Dixmier and Malliavin [16], φ is a finite linear combination of convolutions $\alpha * \beta$ with $\alpha, \beta \in C_c^\infty(G)$. Therefore, $*\varphi$ is a finite linear combination of products of two Hilbert–Schmidt operators and hence is of trace class (as recalled in 9.4).

Remarks. (a) A fastidious reader may have noticed that one point has been glossed over: namely, why is the function $(x, y) \mapsto k_x(y)$ on $\Gamma\backslash G \times \Gamma\backslash G$ measurable? This point is often taken for granted, but is dealt with in [40, §XII.3].

(b) Because 9.3 is valid for $\varphi \in C_c^1(G)$, the previous proof also shows that $*\varphi$ is Hilbert–Schmidt on $°L^2(\Gamma\backslash G)$ already for $\varphi \in C_c^1(G)$. This is established in [41, p. 41].

(c) We have invoked [16] to go from Hilbert–Schmidt to trace class. It is somewhat more elementary to show that, given $n > 0$, a function $\varphi \in C_c^\infty(G)$ is a finite sum of convolutions $\alpha * \beta$ with $\alpha \in C_c^\infty(G)$ and $\beta \in C_c^n(G)$; this follows from the theory of elliptic operators (see [17, p. 199]). Of course, one can bypass the latter by restricting oneself a priori to φ of the form $\alpha * \beta$ with $\alpha, \beta \in C_c^\infty(G)$, as is done in [21]; this is usually sufficient for the applications.

9.6 The remainder of this section is devoted to a generalization of 9.2 and 9.3 that will be needed in Section 11.

Let $\pi: G \rightarrow \Gamma\backslash G$ be the canonical projection. We fix a compact connected subset $C \subset \Gamma\backslash G$ such that $\Gamma\backslash G$ is a finite union of C and of disjoint cusps with smooth boundary (3.18). More precisely, let $\mathcal{P}$ be a set of representatives for the Γ-conjugacy classes of parabolic subgroups P that are cuspidal for Γ. For $P \in \mathcal{P}$, let $\mathfrak{S}_{P,t}$ be a Siegel set with t big enough so that the $\pi(\mathfrak{S}_{P,t})$ are disjoint cusps, and take for C the complement of the union of the $\mathfrak{S}_{P,t}$. We let $°L^2(\Gamma\backslash G, C)$ be the space of elements in $L^2(\Gamma\backslash G)$ that are cuspidal outside C, that is, such that $f_P(x) = 0$ for $x \in \mathfrak{S}_{P,t}$ $(P \in \mathcal{P})$. This space obviously contains $°L^2(\Gamma\backslash G)$.

9.7 Theorem.   *We retain the previous notation.*

(i) *The subspace $°L^2(\Gamma\backslash G, C)$ is closed in $L^2(\Gamma\backslash G)$.*

(ii) *Let $\varphi \in C_c^\infty(G)$ and $D \in \mathcal{U}(\mathfrak{g})$. Then there exists a constant $c(\varphi, D)$ such that*

(1) $|D(f * \varphi)(x)| \leqslant c(\varphi, D)\|f\|_2$ $(f \in °L^2(\Gamma\backslash G, C),\ x \in \Gamma\backslash G)$.

(iii) *The operator $*\varphi\colon °L^2(\Gamma\backslash G, C) \to L^2(\Gamma\backslash G)$ is compact.*

The proof of (i) is the same as that of 8.2: the space $°L^2(\Gamma\backslash G, C)$ is again the intersection of the kernel of linear forms $\lambda_{P,\varphi}$, where φ runs through the elements of $C_c^\infty(G)$ with support outside C. These are among the linear forms considered in 8.2, so they are continuous.

Proof of (ii). There exists a compact connected subset C' of G such that

$$\pi(C') = C.$$

Let U be a relatively compact connected symmetric neighborhood of 1 in G containing $\operatorname{supp}\varphi$. We may choose a compact set $B \supset C$ in $\Gamma\backslash G$ that contains $\pi(C'.U)$. We may assume that it is the complement of the union of Siegel sets $\mathfrak{S}_{P,t'}$ for some $t' > t$ $(P \in \mathcal{P})$.

Any compact set C belongs to a Siegel set and so $a(x)^\rho$ is bounded on C. Therefore 9.3(1) shows the existence of a constant $c_1(\varphi, D) > 0$ such that

(2) $|D(f * \varphi)(x)| \leqslant c_1(\varphi, D)\|f\|_2$ $\big(f \in °L^2(\Gamma\backslash G),\ D \in \mathcal{U}(\mathfrak{g}),\ x \in B\big)$.

Let $x \in \mathfrak{S}_{P,t'}$ and $y \in U$; then $x.y \in \mathfrak{S}_{P,t'}$ because the cusps are the connected components of the complement of B. Therefore, if f is cuspidal outside C then $f_P(xy) = 0$ for $x \in \mathfrak{S}_{P,t'}$ and $y \in U$. As a consequence, $f * \varphi$ is cuspidal on $\mathfrak{S}_{P,t'}$, and is also of uniform moderate growth (5.2). We may therefore use 7.4 and deduce by its repeated application (as in 9.2) the existence of $c_2(\varphi, D)$ such that

(3) $|D(f * \varphi)(x)| \leqslant c_2(\varphi, D)\|f\|_2$ $(x \notin B)$.

Then (ii) follows from (2) and (3).

Now (iii) follows from (i) and (ii) by Ascoli's theorem, as in 9.3.

9.8 Theorem.   *Suppose that Q is the orthogonal projection of $L^2(\Gamma\backslash G)$ onto $°L^2(\Gamma\backslash G, C)$. Then $(*\varphi) \circ Q\colon L^2(\Gamma\backslash G) \to L^2(\Gamma\backslash G)$ is a compact operator.*

The projection Q is a continuous operator. Therefore 9.7(1) is also valid for $(*\varphi) \circ Q$; that is, given $\varphi \in C_c^\infty(G)$ and $D \in \mathcal{U}(\mathfrak{g})$, there exists a constant $d(\varphi, D) > 0$ such that

(1) $\qquad |D(Qf * \varphi)(x)| \leqslant d(\varphi, D)\|f\|_2 \quad (f \in L^2(\Gamma \backslash G),\ x \in \Gamma \backslash G).$

This implies the theorem by the argument given in 9.3.

10 Definition and convergence of Eisenstein series

10.1   We use the notation of Section 2 (2.6–2.8) and Section 5 and first prove the properties (0) to (6) of the function h_P and of the norm defined in 2.8; (P, A) is a cuspidal p-pair and $\mathfrak{S} = \mathfrak{S}_{t,w}$ is a Siegel set with respect to P (3.15).

Let C be a compact set in G. Then

(0)
$$\|gv\| \asymp \|v\| \quad (g \in C, \ v \in \mathbb{R}^2).$$

In fact, it suffices to show that

$$\|gv\| \asymp 1$$

when $g \in G$ and $v \in S^1$. This follows because $C \cdot S^1$ is compact and does not contain the origin.

(1)
$$\|x^{-1}v\| \asymp \|a(x)^{-1}v\| \quad \text{for } x \in \mathfrak{S}, \ v \in \mathbb{R}^2.$$

Writing $x = na(x)k$ as usual, we have

$$\|x^{-1}v\| = \|k^{-1}a(x)^{-1}n^{-1}v\| = \|a(x)^{-1}n^{-1}v\| = \|a(x)^{-1}n^{-1}a(x)a(x)^{-1}v\|.$$

Since $a(x)^{-1}.n(x)a(x)$ is contained in a compact set if $x \in \mathfrak{S}$ (see 3.15(c)), our assertion now follows from (0).

(2)
$$h_P(xy) \asymp h_P(x) \quad \text{for } x \in G, \ y \in C.$$

By definition, this means that

$$\|y^{-1}x^{-1}e_P\| \asymp \|x^{-1}e_P\|,$$

which follows from (0).

Next, we claim

(3)
$$\|x^{-1}y^{-1}e_P\| \succ \|y^{-1}e_P\|.\|x^{-1}e_P\| \quad \text{for all } x \in \mathfrak{S}, \ y \in G;$$

that is,

$$h_P(yx) \prec h_P(y)h_P(x) \quad (y \in G, \ x \in \mathfrak{S}).$$

Let

$$y^{-1}e_P = c_1(y)e_P + c_2(y)\tilde{e}_P,$$

where $\tilde{e}_P$ is a unit vector orthogonal to e_P. Then

$$\|y^{-1}e_P\|^2 = c_1(y)^2 + c_2(y)^2.$$

By (1),

$$\|x^{-1}y^{-1}e_P\| \asymp \|a(x)^{-1}y^{-1}e_P\|$$
$$a(x)^{-1}y^{-1}e_P = h_P(x)^{-1}\big(c_1(y)e_P + h_P^2(x)c_2(y)\tilde{e}_P\big)$$
$$\|a(x)^{-1}y^{-1}e_P\|^2 = h_P(x)^{-2}\big(c_1(y)^2 + h_P^4(x)c_2(y)^2\big)$$
$$\|a(x)^{-1}y^{-1}e_P\|^2 \succ h_P(x)^{-2}(c_1(y)^2 + c_2(y)^2) = h_P(x)^{-2}h_P(y)^{-2}$$

and (3) follows.

(4) $$h_P(\gamma) \prec 1 \quad \text{for } \gamma \in \Gamma.$$

We may assume that (P, A) is the standard p-pair (P_0, A_0) – in other words, that ∞ is a cuspidal point for Γ. If $\gamma \in \Gamma_\infty$ then $a(\gamma) = \pm \text{Id}$ and $h_P(\gamma) = 1$. By 3.7 there exists $r > 0$ such that the lower left entry $c(\gamma)$ of γ is greater than or equal to r in absolute value if $\gamma \notin \Gamma - \Gamma_\infty$. Then (4) follows from 2.8(9).

(5) $$h_P(\gamma cxd) \prec h_P(x) \quad \text{for } \gamma \in \Gamma, \ c, d \in C, \ x \in \mathfrak{S}.$$

This follows from (2), (3), and (4):

$$h_P(\gamma cxd) \asymp h_P(\gamma cx) \prec h_P(\gamma c)h_P(x) \asymp h_P(\gamma)h_P(x) \prec h_P(x).$$

Finally, we show

(6)
$$\begin{array}{c} \text{if } (P', A') \text{ is another cuspidal pair then} \\ h_P(yx) \prec h_P(y)h_{P'}(x) \text{ for } y \in G \text{ and } x \in \mathfrak{S}_{t'} = \mathfrak{S}', \end{array}$$

where $\mathfrak{S}'$ is a Siegel set with respect to P'.

There exists $c \in K$ such that $^c P' = P$, whence $ce_{P'} = e_P$ (up to sign, but that does not matter). Using (3), for $x \in \mathfrak{S}'$ and $y \in G$ we have

$$\|x^{-1}y^{-1}e_P\| = \|x^{-1}y^{-1}ce_{P'}\| \succ \|x^{-1}e_{P'}\|\|y^{-1}ce_{P'}\| = \|x^{-1}e_{P'}\|\|y^{-1}e_P\|,$$

which implies (6) in view of the definitions (2.8).

10.2   Assume that (P, A) is a cuspidal p-pair for Γ. As in 3.6, let

$$\Gamma_N = \Gamma \cap N, \qquad \Gamma_P = \Gamma \cap P.$$

We recall that

(1) $$\Gamma_P \subset CG.\Gamma_N, \quad [\Gamma_P : \Gamma_N] \leqslant 2.$$

Let

(2) $$F(P, m) = \{\varphi \in C^\infty(\Gamma_P.N.A\backslash G) \mid \varphi \text{ of right } K\text{-type } m\}.$$

In particular,

(3) $$\varphi(nak) = \varphi(1)\chi_m(k) \quad (n \in N, \, a \in A, \, k \in K);$$

hence dim $F(P, m) \leqslant 1$. Assume that $\Gamma_P \neq \Gamma_N$. Then Γ_P contains an element $\varepsilon.n_o$ $(n_o \in N)$, where $\varepsilon = -\,\mathrm{Id}$ is the nontrivial element of CG and

$$\Gamma_P = \Gamma_N \cup \varepsilon n_o.\Gamma_N, \quad \Gamma_P.N = N.CG, \quad [\,\Gamma_P : \Gamma_N\,] = 2.$$

(If $n_o = 1$ then $\Gamma_P = \Gamma_N \times CG$; if $n_o \neq 1$ then Γ_P is infinite cyclic generated by $\varepsilon.n_o$.) Since ε is central and belongs to K, we have

$$\varphi(x) = \varphi(\varepsilon x) = \varphi(nx\varepsilon) = \varphi(x)\chi_m(\varepsilon) \quad (x \in G).$$

Therefore, if $\chi_m(\varepsilon) = -1$, the function φ is zero and dim $F(P, m) = 0$. *In the sequel, we assume that either $\Gamma_P = \Gamma_N$* (which is equivalent to requiring that Γ_P be "neat"; see 10.12) *or m is even*. Then dim $F(P, m) = 1$ and we let $\varphi_{P,m}$, or simply φ_P if m is clear from the context, be the element of $F(P, m)$ equal to 1 at the identity. Then

(4) $$\varphi_P(nak) = \chi_m(k) \quad (n \in N, \, a \in A, \, k \in K).$$

Recall that $P = NA \cup \varepsilon.NA$. The element ε is central and belongs to K, so

(5) $$\varphi_P(px) = \varphi_P(p).\varphi_P(x) \quad (p \in P, \, x \in G),$$

(6) $$\varphi_P(p) = \begin{cases} 1 & \text{if } p \in NA, \\ (-1)^m & \text{if } p \in \varepsilon.NA. \end{cases}$$

In the sequel, we let

(7) $$\varphi_{P,s,m} = \varphi_{P,s} = \varphi_P.h_P^{1+s}.$$

The properties of φ_P and h_P (see (5) and 2.8(4)) imply that

(8) $$\varphi_{P,s}(paxk) = \varphi_P(p).h_P(a)^{1+s}.\varphi_{P,s}(x)\chi_m(k)$$

$(p \in P, \, a \in A, \, x \in G, \, k \in K)$. Note also that *if m is even then $\varphi_{P,s}$ is right- and left-invariant under CG*.

Let $k \in K$ and $P' = {}^kP$. Then $A' = {}^kA$ (recall we consider only normal p-pairs (2.6)). It is readily checked that

(9) $$\varphi_{P'}({}^kx) = \varphi_P(x) \quad (x \in G),$$

hence also, in view of 2.8(5),

(10) $$\varphi_{P',s}({}^kx) = \varphi_{P,s}(x) \quad (k \in K, \, P' = {}^kP, \, x \in G).$$

10.3 Definition. The *Eisenstein series* with respect to P and $s \in \mathbb{C}$ is

(1) $$E(P, s)(g) = \sum_{\gamma \in \Gamma_P \backslash \Gamma} \varphi_{P,s}(\gamma g).$$

Here again, m is understood. If it needs to be made explicit, we shall write $E(P, m, s)$. If m is odd and $\Gamma_P \neq \Gamma_N$, then $E(P, m, s) \equiv 0$ by definition.

In the standard situation where $P = P_0$, we have

$$h_P(\gamma x) = \mathcal{I}(\gamma x(i))^{1/2} = \frac{(\mathcal{I}x(i))^{1/2}}{|\mu(\gamma, x(i))|};$$

hence

$$(2) \qquad E(P_0, s)(x) = \sum_{\gamma \in \Gamma_{P_0} \backslash \Gamma} \frac{y^{(s+1)/2} \varphi_P(\gamma x)}{|c_\gamma x(i) + d_\gamma|^{s+1}} \qquad \left(\gamma = \begin{pmatrix} a_\gamma & b_\gamma \\ c_\gamma & d_\gamma \end{pmatrix} \right).$$

If $m = 0$, then E is invariant under K and may be viewed as a function on X, which can be written

$$(3) \qquad \sum \frac{y^{(s+1)/2}}{|c_\gamma z + d_\gamma|^{s+1}} = \sum \frac{y^{(s+1)/2}}{|\mu(\gamma, z)|^{s+1}} \qquad (z = x(i)).$$

Returning to the general case, we write s in the form $s = \sigma + i\tau$ ($\sigma, \tau \in \mathbb{R}$) and denote by E_0 the termwise absolute majorant of $E(P, s)$:

$$(4) \qquad E_0(g) = \sum_{\gamma \in \Gamma_P \backslash \Gamma} |\varphi h_P^{1+s}(\gamma g)| = \sum_{\gamma \in \Gamma_P \backslash \Gamma} h_P(\gamma g)^{1+\sigma}.$$

Note that E_0 is independent of the K-type m.

10.4 Theorem.

(i) $E = E(P, s)$ *converges absolutely and locally uniformly for $\sigma > 1$.*

(ii) *If (P', A') is a cuspidal pair and $\mathfrak{S}' = \mathfrak{S}_{P',t}$ a Siegel set with respect to (P', A'), then there exists $c > 1$ such that*

$$E_0(x) \prec \frac{c^\sigma}{\sigma - 1} h_{P'}^{\sigma+1}(x) \quad \text{for } x \in \mathfrak{S}', \ \sigma > 1.$$

(iii) $\mathcal{C}E = ((s^2 - 1)/2)E$ *for $\sigma > 1$, where $\mathcal{C}$ is the Casimir operator.*

(iv) E *is an automorphic form for $\sigma > 1$.*

(v) *If P' is a cuspidal subgroup for Γ not conjugate to P by Γ, and if $\mathfrak{S}'$ is a Siegel set with respect to P', then*

$$E_0(x).h_{P'}(x)^{-(\sigma+1)} \to 0 \quad \text{on } \mathfrak{S}'$$

when $h_{P'}(x) \to \infty$.

Proof. Note that (ii) implies (i), since any compact set is contained in a Siegel set; (ii) also shows that E has moderate growth. Because $\varphi_{P,s}$ is of type m, the series E is obviously of type m; of course, E is left-invariant for Γ. Hence there remains to prove (ii), (iii), and (v). We give Godement's proof of (ii) first.

By 10.1(6), we have

$$h_P(yx) \prec h_P(y)h_{P'}(x) \quad (y \in G, \ x \in \mathfrak{S}');$$

therefore,

(1) $\quad E_0(x) = \sum h_P(\gamma x)^{1+\sigma} \prec h_{P'}(x)^{1+\sigma} \sum h_P(\gamma)^{1+\sigma} \quad (x \in \mathfrak{S}', \ \gamma \in \Gamma).$

As a consequence, to prove (ii) it suffices to show the existence of $c > 1$ such that

(2) $\qquad \sum h_P(\gamma)^{\sigma+1} \prec c^{\sigma}(\sigma - 1)^{-1} \quad \text{for } \sigma > 1.$

Fix a symmetric compact neighborhood C of 1 such that $\Gamma \cap C.C = \{1\}$. Then

$$\gamma C \cap \gamma' C \neq \emptyset \Rightarrow \gamma = \gamma' \quad (\gamma, \gamma' \in \Gamma).$$

By 10.1(2) and (4) we can find constants $b, d > 0$ such that

(3) $\qquad h_P(\gamma) \leqslant d.h_P(\gamma x) \leqslant d.b \quad (\gamma \in \Gamma, \ x \in C).$

Raising to the power $\sigma + 1$ and integrating over C we obtain, from (3) and (1),

(4) $\qquad h_P(\gamma)^{\sigma+1} \leqslant \mu(C)^{-1} d^{\sigma+1} \int_C h_P(\gamma x)^{\sigma+1} \, dx,$

(5) $\quad E_0(x) \leqslant d^{\sigma+1} \mu(C)^{-1} \left(\sum_{\gamma \in \Gamma_P \backslash \Gamma} \int_{\gamma C} h_P(y)^{\sigma+1} \, dy \right) h_{P'}(x)^{\sigma+1} \quad (x \in \mathfrak{S}').$

Let

$$A(0, b) = \{ a \in A \mid h_P(a) \leqslant b \} \quad (b > 0).$$

Then, by (3),

(6) $\qquad\qquad\qquad \Gamma C \subset NA(0, d)K;$

hence

$$\Gamma_P \backslash \Gamma C \subset \Gamma_N \backslash N \times A(0, d) \times ((CG \cap \Gamma) \backslash K).$$

The Haar measure on G can be written

$$dg = a(g)^{-2} dn.da.dk,$$

as was recalled in 2.9. On A we take $h = h_P(a) = a^{\rho}$ as coordinate (see 2.8). Then the restriction to A of the Haar measure on G is $dh.h^{-3}$. Therefore, (5) and (6) imply the existence of a constant $\delta > 0$ such that

(7) $\qquad E_0(x) \leqslant \delta \left(\int_{\Gamma_P \backslash NK} dn \, dk \right) \left(\int_0^d h^{\sigma-2} \, dh \right) d^{\sigma+1} h_{P'}^{\sigma+1}(x).$

The first integral is finite and strictly positive. The second one converges for $\sigma > 1$ and is then equal to $d^{\sigma-1}(\sigma - 1)^{-1}$. Finally,

$$E_0(x) \leqslant \delta_1 \frac{c^{\sigma-1}}{(\sigma-1)} h_{P'}^{\sigma+1}(x) \quad (x \in \mathfrak{S}')$$

for some constant $\delta_1 > 0$, with $c = d^2$. The right-hand side increases if c is replaced by a bigger constant, so we may assume $c > 1$.

To establish (iii), it suffices to show that

(8) $$C\varphi_{P,s} = \frac{s^2-1}{2}\varphi_{P,s}.$$

The function h_P is N-invariant on the left and K-invariant on the right, as follows from its definition (2.8). In view of 10.2(8), we have

$$\varphi_{P,s}(nak) = h_P^{s+1}(a)\chi_m(k) \quad (n \in N, \, a \in A, \, k \in K).$$

Because C is bi-invariant, $C\varphi_{P,s}$ is also left N-invariant and of right K-type m, so it suffices to check (8) for $x \in A$. After conjugation, we may assume that the cuspidal point is ∞ and that $P = P_0$. Then (see 2.5(1))

$$C = \tfrac{1}{2}H^2 + EF + FE = \tfrac{1}{2}H^2 - H + 2EF.$$

Since A normalizes N, we have

$$\varphi_{P,s}(ae^{t_1 E}e^{t_2 F}) = (\varphi_{P,s})(ae^{t_2 F}) \quad (a \in A);$$

therefore (see 2.1(6)),

(9) $$EF(\varphi_{P,s}) = 0,$$

so that

$$C\varphi_{P,s}(a) = \left(\tfrac{1}{2}H^2 - H\right)\varphi_{P,s}(a) \quad (a \in A).$$

On A, $\varphi_{P,s} = h_P^{1+s}$. Take again h_P as coordinate on A and identify A with its Lie algebra $\mathfrak{a} \cong \mathbb{R}$ by the map $a \mapsto \log h_P(a)$. Then $h_P^{s+1}(a) = e^{t(s+1)}$, where $e^t = a$, and H becomes d/dt. Because

$$\left(\frac{1}{2}\frac{d^2}{dt^2} - \frac{d}{dt}\right)e^{t(s+1)} = \left(\frac{(s+1)^2}{2} - (s+1)\right)e^{t(s+1)} = \left(\frac{s^2-1}{2}\right)e^{t(s+1)},$$

(iii) is proved.

Proof of (v). By 10.1(6), there exists a constant $c_1 > 0$ such that

$$h_P(\gamma x)^{1+\sigma}.h_{P'}(x)^{-(\sigma+1)} \leqslant c_1 h_P(\gamma)^{\sigma+1} \quad (\gamma \in \Gamma, \, x \in \mathfrak{S}').$$

By (ii), the series

$$\sum_{\Gamma_P \backslash \Gamma} h_P(\gamma)^{\sigma+1}$$

is convergent for $\sigma > 1$, so it is enough to show that, for a given $\gamma \in \Gamma$,

$$h_P(\gamma x).h_{P'}(x)^{-1} \to 0 \quad \text{as } h_{P'}(x) \to \infty \quad (x \in \mathfrak{S}').$$

We claim first that it suffices to prove

(10) $$h_P(\gamma x) \asymp h_{P'}(x)^{-1} \quad (x \in \mathfrak{S}').$$

Indeed, (10) implies that

$$h_P(\gamma x).h_{P'}(x)^{-1} \asymp h_{P'}(x)^{-2} \quad (x \in \mathfrak{S}')$$

and the last term clearly tends to zero if $h_{P'}(x) \to \infty$. There remains to prove (10). We can write

(11) $$\gamma^{-1}e_P = \lambda e_{P'} + \mu \tilde{e}_{P'} \quad (\lambda, \mu \in \mathbb{R}).$$

We claim that $\mu \neq 0$. Assume to the contrary that $\mu = 0$. Then we have $\gamma^{-1}e_P = \lambda e_{P'}$, which implies that $^\gamma P' = P$, in contradiction with the assumption that P and P' are not conjugate under Γ. Therefore $\mu \neq 0$.

Using the Iwasawa decomposition with respect to P', we can write

$$x = n(x).a_{P'}(x).k(x) \quad \left(n(x) \in N', \, a_{P'}(x) \in A_{P'}, \, k(x) \in K\right).$$

Then

$$h_P(\gamma x)^{-1} = \|a_{P'}(x)^{-1}.n(x)^{-1}.\gamma^{-1}e_P\|.$$

But $n(x)^{-1}$ is upper triangular with respect to the basis $(e_{P'}, \tilde{e}_{P'})$ and unipotent, so

$$n(x)^{-1}.e_{P'} = e_{P'}, \quad n(x)^{-1}.\tilde{e}_{P'} = \tilde{e}_{P'} + \nu(x).e_{P'} \quad (\nu(x) \in \mathbb{R})$$

and, using (11), we obtain

$$\|a_{P'}(x)^{-1}n(x)^{-1}(\lambda e_{P'} + \mu \tilde{e}_{P'})\| = \|a_{P'}(x)^{-1}(\lambda'(x)e_{P'} + \mu \tilde{e}_{P'})\|,$$

where $\lambda'(x) = \lambda + \nu(x)$ is bounded in absolute value (recall that $n(x) \in \omega$).

From

$$a_{P'}(x)e_{P'} = h_{P'}(x)e_{P'}, \qquad a_{P'}(x)\tilde{e}_{P'} = h_{P'}(x)^{-1}\tilde{e}_{P'}$$

(see 2.8), we have

$$\begin{aligned} h_P(\gamma x)^{-1} &= \|\lambda'(x)h_{P'}(x)^{-1}e_{P'} + \mu h_{P'}(x)\tilde{e}_{P'}\| \\ &= h_{P'}(x)\|\lambda'(x)h_{P'}(x)^{-2}e_{P'} + \mu \tilde{e}_{P'}\| \\ &= h_{P'}(x)\left(\lambda'(x)^2 h_{P'}(x)^{-4} + \mu^2\right)^{1/2}. \end{aligned}$$

We have already shown that $\mu \neq 0$. Since $\lambda'(x)^2$ is bounded and $h_{P'}(x)^{-1} \prec 1$ on $\mathfrak{S}'$, the second factor in the right-hand side is $\asymp 1$, and (10) follows.

10.5 Proposition. *Let $\alpha \in I_c^\infty(G)$ (see 2.12(2)), and let P be a parabolic subgroup. Then the function on $\mathbb{C}$ defined by*

(1) $$\pi_\alpha(s) = \int_G \varphi_{P,s}(g)\alpha(g^{-1})\,dg$$

is entire, independent of P, of right K-type m, and satisfies the relations

(2) $$\varphi_{P,s} * \alpha = \pi_\alpha(s)\varphi_{P,s} \quad (s \in \mathbb{C}),$$

(3) $$E(P, s) * \alpha = \pi_\alpha(s)E(P, s) \quad (\mathcal{R}s = \sigma > 1).$$

Proof. The first assertion follows since $\varphi_{P,s}$ is (obviously) entire in s of right K-type m and α is K-invariant. Let now (P', A') be another normal p-pair and N' the unipotent radical of P'. There exists $k \in K$ such that ${}^kP = P'$ and ${}^kA = A'$. By 10.2(9),

$$\varphi_{P',s}({}^kx) = \varphi_{P,s}(x) \quad (x \in G).$$

Because α is K-invariant, we have $\alpha({}^kx) = \alpha(x)$ $(x \in G)$. Moreover, $d({}^kx) = dx$ since the Haar measure is bi-invariant. Therefore, replacing P by P' in (1) does not change the value of the left-hand side, which shows that $\pi_\alpha(s)$ does not depend on P.

Because α is K-invariant and $\varphi_{P,s}$ is of K-type m on the right, it follows that $\varphi_{P,s} * \alpha$ is also of right K-type m. It suffices to check (2) for $x = na$ $(n \in N, a \in A)$:

$$(\varphi_{P,s} * \alpha)(na) = \int_G \varphi_P(nag)h_P(nag)^{s+1}\alpha(g^{-1})\,dg.$$

But φ_P is left-invariant under NA and $h_P(nag) = h_P(a)h_P(g)$ (see 2.8), so

$$(\varphi_{P,s} * \alpha)(na) = h_P(a)^{s+1}\int_G \varphi_P(g)h_P(g)^{s+1}\alpha(g^{-1})\,dg,$$
$$(\varphi_{P,s} * \alpha)(na) = h_P(a)^{s+1}\pi_\alpha(s).$$

As $\varphi_P \equiv 1$ on NA, this proves (2). By the absolute and locally uniform convergence, we can permute integration and summation, whence (3).

The following simple remark will play an important role in Section 11.

10.6 Lemma. *Given $c \in \mathbb{C}$, there exists a neighborhood U of c and $\alpha \in I_c^\infty(G)$ such that*

(1) $$|\pi_\alpha(s)| \geqslant 1/2 \quad (s \in U).$$

Let $\{\alpha_n\}$ $(n = 1, 2, \dots)$ be a Dirac sequence in $I_c^\infty(G)$ (see 2.4). Then

$$\varphi_{P,s} * \alpha_n \to \varphi_{P,s}$$

(uniformly on compact sets, by 2.4). Since

$$\varphi_{P,s} * \alpha_n = \pi_{\alpha_n}(s)\varphi_{P,s}$$

by 10.5, and since $\varphi_{P,s} \neq 0$ (in fact, is nowhere zero), this shows the existence of $\alpha \in I_c^\infty(G)$ such that $\pi_\alpha(c) \neq 0$. We may therefore find α such that $\pi_\alpha(c) = 1$, and then the existence of U is clear.

10.7 Definition. Given $c \in \mathbb{C}$, a neighborhood U of c in $\mathbb{C}$ and $\alpha \in I_c^\infty(G)$ are said to be *compatible* with one another if they satisfy 10.6(1).

10.8 Eisenstein series on the upper half-plane. We relate the Eisenstein series considered previously with some of the classical ones on X.

Consider the case where $m = 0$ (i.e., where $\varphi_P \equiv 1$) and the corresponding Eisenstein series (10.3), which is now K-invariant on the right and hence can be viewed as a function on X – namely,

$$(1) \qquad E(P, s)(z) = \sum_{\gamma \in \Gamma_P \backslash \Gamma} \frac{y^{(s+1)/2}}{|\mu(\gamma, z)|^{s+1}} \qquad (z \in \mathsf{X});$$

it is Γ-invariant on the left. A procedure inverse to the one of 5.13 leads us to associate to it the function

$$(2) \qquad \tilde{E}(P, s)(z) = \sum_{\gamma \in \Gamma_P \backslash \Gamma} |\mu(\gamma, z)|^{-(s+1)} \qquad (z \in \mathsf{X}),$$

which is now automorphic for Γ with automorphy factor $|\mu(\gamma, z)|^{s+1}$; that is,

$$(3) \qquad \tilde{E}(P, s)(\gamma z) = |\mu(\gamma, z)|^{s+1} \tilde{E}(P, s)(z),$$

as follows from the cocycle identity 3.3(7). If now $s = m \in \mathbb{N}$, we may dispense with the absolute value and define

$$(4) \qquad E(P, s) = \sum_{\gamma \in \Gamma_P \backslash \Gamma} \mu(\gamma, z)^{-m},$$

which converges for $m \geqslant 3$. It is identically zero if $\Gamma_P \neq \Gamma_N$ and m is odd (see 10.2). In the sequel we rule out this case. Note that if $-1 \in \Gamma$ and m is even then

$$\sum_{\Gamma_N \backslash \Gamma} \mu(\gamma, z)^{-m} = 2E(P, s).$$

Assume now that $\Gamma = \mathrm{SL}_2(\mathbb{Z})$ and therefore that $m = 2k$ is even. Write e_k for $E(P_0, 2k)$. We have

$$e_k(z) = \sum_{\substack{c,d\in\mathbb{Z} \\ (c,d)=1 \\ (c,d)\neq(0,0)}} (cz+d)^{-2k}.$$

Let

$$G_k(z) = \sum_{\substack{(m,n\in\mathbb{Z}) \\ (m,n)\neq(0,0)}} (mz+n)^{-2k}.$$

Clearly

$$G_k(z) = \zeta(2k).e_k(z).$$

The G_k are the original Eisenstein series. Let $\omega_1, \omega_2 \in \mathbb{C}^*$ be such that

$$z = \omega_1/\omega_2 \in X.$$

Then

$$\sum_{\substack{m,n\in\mathbb{Z} \\ (m,n)\neq(0,0)}} (m\omega_1 + n\omega_2)^{-2k} = \omega_2^{-2k} \sum_{\substack{m,n\in\mathbb{Z} \\ (m,n)\neq(0,0)}} (mz+n)^{-2k}.$$

Essentially, e_k is a function on lattices (i.e., on elliptic curves), with $z = \omega_1/\omega_2$ identified with $\mathbb{C}/(\mathbb{Z}\omega_1 + \mathbb{Z}\omega_2)$. For all this see for example [52] or [53].

10.9 Haar measures. We shall need a slight generalization of 2.9(b) for $H = G$. Let $L \supset M$ denote closed unimodular subgroups of G. Choose Haar measures $d\mu_G, d\mu_L, d\mu_M$ on G, L, M. Let $d\mu_{M\backslash G}, d\mu_{M\backslash L}, d\mu_{L\backslash G}$ be the quotient Haar measures on $M\backslash G, M\backslash L, L\backslash G$ as characterized in 2.9(b). Then $d\mu_{L\backslash G}$ is a quotient of $d\mu_{M\backslash G}$ by $d\mu_{M\backslash L}$ in the following sense: if f is continuous and is integrable on $M\backslash H$, then

$$h \mapsto \int_{M\backslash L} |f(xh)| \, d\mu_{M\backslash L}(x)$$

is integrable on $L\backslash H$ and

(1) $$\int_{M\backslash H} f(x) \, d\mu_{M\backslash L}(x) = \int_{L\backslash H} d\mu_{L\backslash H}(\dot{x}) \int_{M\backslash L} f(yx) \, d\mu_{M\backslash L}(y).$$

Conversely, let f be a continuous function on $M\backslash H$. Assume that

$$x \mapsto \int_{M\backslash L} |f(yx)| \, d\mu_{M\backslash L}(y)$$

is integrable on $L\backslash H$. Then f is integrable on $M\backslash H$ and the integral is therefore given by (1). For all this (in much greater generality, needless to say), see [10, §VII.8, p. 64, Cor. to Prop. 12].

If f and g are functions on $L\backslash H$, we let

(2) $$(f, g)_{L\backslash H} = \int_{L\backslash H} f(x)\overline{g(x)}\, d\mu_{L\backslash H}(x).$$

10.10 Proposition. *Let f be a continuous, fast decreasing function on $\Gamma\backslash G$. Then, for $\mathcal{R}s > 1$,*

(1) $$(E(P, s), f)_{\Gamma\backslash G} = (\varphi_{P,s}, f_P)_{N\backslash G}.$$

Proof. The function $\varphi_{P,s}$ is left-invariant under Γ_P. We show first that $\varphi_{P,s}.\bar{f}$ is integrable on $\Gamma_P\backslash G$ and that

(2) $$(E(P, s), f)_{\Gamma\backslash G} = (\varphi_{P,s}, f)_{\Gamma_P\backslash G}.$$

We want to apply 10.9(1) to the case where $H = G$, $L = \Gamma$, and $M = \Gamma_P$. Note first that

$$\int_{\Gamma_P\backslash\Gamma} \bar{f}(\gamma x).\varphi_{P,s}(\gamma x)\, d\mu_{\Gamma_P\backslash\Gamma} = \sum_{\gamma\in\Gamma_P\backslash\Gamma} \bar{f}(\gamma x).\varphi_{P,s}(\gamma x)$$

(3) $$= \bar{f}(x).E(P, s)(x) \quad (x \in G).$$

In the proof of 10.4, we have seen that

$$\sum_{\gamma\in\Gamma_P\backslash\Gamma} |\varphi_{P,s}(\gamma x)|$$

is of moderate growth; hence so is $|E(P, s)|$. Its product with $\bar{f}$ is then fast decreasing, so $E(P, s).\bar{f}$ is integrable on $\Gamma\backslash G$. Therefore (see 10.9), $\varphi_{P,s}.\bar{f}$ is integrable on $\Gamma\backslash G$ and so 10.9(1) yields (2). To prove (1), it suffices to show that

(4) $$(\varphi_{P,s}, f)_{\Gamma_P\backslash G} = (\varphi_{P,s}, f_P)_{N\backslash G}.$$

The relation 10.9(1), when applied to $H = G$, $L = \Gamma_P.N$, and $M = N$, yields

(5) $$(\varphi_{P,s}, f)_{\Gamma_P\backslash G} = \int_{\Gamma_P.N\backslash G} \varphi_{P,s}(\dot{x})\, d\dot{x} \int_{\Gamma_P\backslash\Gamma_P.N} \bar{f}(n.\dot{x})\, dn.$$

We claim that

(6) $$\int_{\Gamma_P N\backslash G} \varphi_{P,s}(\dot{x})\, d\dot{x} \int_{\Gamma_P\backslash\Gamma_P.N} \bar{f}(n.\dot{x})\, dn$$

$$= \int_{N\backslash G} \varphi_{P,s}(\dot{x})\, d\dot{x} \int_{\Gamma_N\backslash N} \bar{f}(n.\dot{x})\, dn.$$

If $\Gamma_P = \Gamma_N$ then this is obvious, so we assume $\Gamma_P \neq \Gamma_N$. Then (see 10.2)

$$\Gamma_P\backslash\Gamma_P N = \Gamma_N\backslash N \cup \varepsilon n_\circ \Gamma_N\backslash\varepsilon N$$

and, as a result,

$$(7) \qquad \int_{\Gamma_P \backslash \Gamma_P N} \bar{f}(n.x)\, dn = 2.\bar{f}_P(x).$$

On the other hand, $N \backslash G$ is a 2-fold covering of $\Gamma_P N \backslash G$; hence

$$(8) \qquad \int_{\Gamma_P N \backslash G} \varphi_{P,s}(\dot{x}) \bar{f}_P(\dot{x})\, d\dot{x} = \frac{1}{2} \int_{N \backslash G} \varphi_{P,s}(\dot{x}) \bar{f}_P(\dot{x})\, d\dot{x}$$

and (6) follows. The right-hand side of (5) is therefore equal to $(\varphi_{P,s}, f_P)_{N \backslash G}$ and (4) is proved.

Remark. The proof of (4) remains valid if $\varphi_{P,s}$ is replaced by a Γ_P-invariant function of moderate growth. The function f was assumed to be fast decreasing to make sure the integrals were convergent, but the proof holds under the weaker assumption.

10.11 Corollary. *For $\mathcal{R}s > 1$, the Eisenstein series $E(P, s)$ is orthogonal to all cusp forms.*

10.12   We now return to the condition $\Gamma_P = \Gamma_N$ considered in 10.2 and the notion of a neat linear group. A subgroup L of $GL_n(\mathbb{R})$ (or $GL_n(F)$, where F is any commutative field) is *neat* if, for each $x \in L$, the subgroup $A(x)$ of $\mathbb{C}^*$ (or of $\bar{F}^*$, where $\bar{F}$ is an algebraic closure of F) generated by the eigenvalues of x is torsion-free [6, 17.1]. If F has characteristic zero and L is finitely generated, then L has a neat subgroup of finite index [6, 17.7]. It is clear from the discussion in 10.2 that $\Gamma_P = \Gamma_N$ if and only if Γ_P is neat. We shall also say that the associated cusp is neat. In the terminology of [53, p. 29], a cusp is regular if neat, irregular otherwise.

As in 9.6, we consider a fundamental set consisting of disjoint Siegel sets $\mathfrak{S}_{i,t}$ and a compact set C, and assume the cusps are parameterized by the equivalence classes of Γ-cuspidal points. We let $(P_1, A_1), \ldots, (P_l, A_l)$ be the corresponding p-pairs. In the sequel, we let $\delta \in \{0, 1\}$, with $l(\delta)$ equal to the number l of cusps if $\delta = 0$ and to the number of neat cusps if $\delta = 1$. Moreover, we assume the cusps to be so numbered that the first $l(\delta)$ are neat.

If m is odd and $i > l(\delta)$, then $F(P_i, m) = 0$ (see 10.2). Therefore:

(1) $E(P_i, m, s) = 0$ if m is odd and $i > l(\delta)$;
(2) if $f \in C(\Gamma \backslash G)$ is of right K-type m and m is odd, then $f_{P_i} = 0$ for $i > l(\delta)$.

11 Analytic continuation of the Eisenstein series

In the right half-plane $\mathcal{R}s > 1$, the Eisenstein series is holomorphic in s. Our main goal is to prove it has a meromorphic continuation to $\mathbb{C}$ that satisfies a functional equation. This was first established by A. Selberg, who described his proof in lectures at Göttingen in 1956; however, these lectures have only recently been published [50]. We shall give J. Bernstein's proof. It has its origin in A. Selberg's so-called third proof, sketched in Appendix F of [33], and also, more recently, in the introduction to [50]. So far, I have only seen brief accounts by Bernstein himself. I have benefited from a more detailed presentation by D. Miličić (unpublished).

We shall use the notation of Sections 9 and 10 rather freely.

11.0 We first make some remarks on the notions of meromorphic function that will occur here. Roughly, it will be shown that there exists a discrete set C in $\mathbb{C}$, contained in $\mathcal{R}s \leqslant 1$, and a function $F(s, x)$ smooth on $(\mathbb{C} - C) \times (\Gamma \backslash G)$, holomorphic in s, that coincides with $E(s, x)$ if $\mathcal{R}s > 1$. Moreover, to each $s \in \mathbb{C}$ there is attached an $m(s) \in \mathbb{N}$, which is zero if $s \notin C$. Given $c \in \mathbb{C}$, there is a disc $D(c, R) = \{ s \in \mathbb{C} \mid |s - c| \leqslant R \}$, so we can write

$$(1) \qquad (s - c)^{m(c)} F(s, x) = \sum_{j \geqslant 0} (s - c)^j F_j(x) \quad (F_0 \neq 0),$$

where the F_j are smooth in x and the series converges absolutely and uniformly on $D(c, R) \times L$ for any $L \subset \Gamma \backslash G$ compact.

Recall that $C^\infty(\Gamma \backslash G)$ is usually viewed as a topological vector space with the topology defined by absolute and uniform convergence on compact sets – that is, by the seminorms $v_{D,L}$ (L compact in $\Gamma \backslash G$, $D \in \mathcal{U}(\mathfrak{g})$) – where

$$v_{D,L}(f) = \max_{x \in L} |Df(x)|$$

(we have used 2.1). The proof will show, in fact, that

$$(2) \qquad \sum_{n \geqslant 0} R^n v_{D,L}(F_n) < \infty$$

for all the seminorms $v_{D,L}$. The latter means that the left-hand side is a holomorphic function with values in the (quasicomplete) topological vector space $C^\infty(\Gamma \backslash G)$ (cf. [13, §3]).

Such a function is *weakly holomorphic*: given a continuous linear form μ on $C^\infty(\Gamma \backslash G)$, the function $s \mapsto (s - c)^m . \mu(F(s, \cdot))$ is holomorphic. Since μ is continuous by definition, given $v_{D,L}$ there exists a constant $c(D, L)$ such that

$$(3) \qquad |\mu(f)| \leqslant c(D, L) . v_{D,L}(f) \quad (f \in C^\infty(\Gamma \backslash G));$$

99

this follows obviously from (2). There is a converse. The previous condition for all continuous linear forms implies holomorphy (see [61, §V.3] or [46, VI.3, Thm. VI.4] for the case of Banach spaces), but we shall not need this result.

In the course of the proof, we will also encounter holomorphic functions with values in a Hilbert space H. A function $s \mapsto G(s) \in H$ is holomorphic in a domain U if we can find around each $c \in U$ a disc $D(c, R)$ and a power series

$$\sum_{j \geqslant 0} (s - c)^j h_j$$

with $h_j \in H$, such that

$$\sum R^j \|h_j\| < \infty$$

and such that, for each $s \in D(c, R)$, the partial sums converge to $G(s)$ in H.

Here again this notion is equivalent to weak holomorphy: for each continuous linear form λ on H, the scalar function $s \mapsto \lambda(G(s))$ is holomorphic.

11.1 Lemma. *Let f be an automorphic form for Γ, of right K-type m. Assume that f is an eigenvector of C with eigenvalue $(s^2 - 1)/2$, where $s \neq 0$. Let P be a Γ-cuspidal parabolic subgroup. Then there exist $\mu, \nu \in F(P, m)$ such that*

(1) $$f_P = \mu h_P^{s+1} + \nu h_P^{1-s} \qquad (s \in \mathbb{C}^*)$$

(cf. 10.2(2) for $F(P, m)$).

The constant term f_P is of K-type m on the right and is N-invariant on the left; it is therefore determined by its restriction to A and, by the same argument as in the proof of 10.4(iii), satisfies the equation

(2) $$(\tfrac{1}{2} H^2 - H) f_P(a) = 2^{-1}(s^2 - 1) f_P(a) \quad (a \in A).$$

The space of solutions of that ordinary differential equation is 2-dimensional. For $s \neq 0$, the equation

(3) $$(\lambda^2/2) - \lambda = (s^2 - 1)/2$$

has two distinct solutions: $1 \pm s$. The space of solutions is therefore spanned by the eigenfunctions for H with eigenvalues $1 \pm s$, which are h_P^{1+s} and h_P^{1-s}; moreover, on A, f_P is a linear combination of those functions with constant coefficients. Both sides are N-invariant on the left. The function f_P is of K-type m on the right, whereas $h_P^{1\pm s}$ is K-invariant on the right. It follows that $\mu, \nu \in F(P, m)$, as claimed.

Hence μ and ν are multiples of $\varphi_{P,m}$ (cf. 10.2), and we can write

(4) $$f_P = d.\varphi_{P,s} + c.\varphi_{P,-s} \quad (c, d \in \mathbb{C}).$$

11.2 We now apply the foregoing to the Eisenstein series. If P and P' are Γ-cuspidal then for $\mathcal{R}s > 1$ we have functions of s, to be denoted $d_{P|P'}(s)$ and $c_{P|P'}(s)$, such that

(1) $$E(P, s)_{P'} = d_{P|P'}(s)\varphi_{P',s} + c_{P|P'}(s)\varphi_{P',-s}.$$

11.3 Proposition.

  (i) $d_{P|P} \equiv 1$.
  (ii) *If P' is not Γ-conjugate to P, then $d_{P|P'}(s) = 0$.*
  (iii) *There exists a constant $C > 1$ such that*

$$|c_{P|P'}(s)| \prec C^{\sigma}.(\sigma - 1)^{-1} \quad (1 < \sigma < \infty, \; \sigma = \mathcal{R}s).$$

Proof. The second assertion follows immediately from 10.4(v): by that assertion, for a given s with $\mathcal{R}s = \sigma > 1$, $E(P, s)(x)/h_{P'}^{s+1}(x) \to 0$ in a Siegel set $\mathfrak{S}'$ for P'. The proof shows that this is uniform on compact subsets of $\Gamma \backslash G$, so we also have

$$E(P, s)_{P'}(x)/h_{P'}^{s+1}(x) \to 0 \quad \text{as } h_{P'}(x) \to \infty \text{ in } \mathfrak{S}'.$$

We then divide both sides of 11.2(1) by $h_{P'}(x)^{s+1}$ and note that, for $\sigma > 1$, both $h_{P'}^{-s-1}$ and $h_{P'}^{1-s}$ tend to zero as $h'_P(x) \to \infty$.

Proof of (i). Using the Bruhat decomposition 2.7(2), we can write

$$\Gamma = \Gamma_P \coprod \Gamma_w \quad (\Gamma_w = \Gamma \cap PwN)$$

and hence

$$\Gamma_P \backslash \Gamma = \{1\} \coprod \Gamma_P \backslash \Gamma_w.$$

The sum defining the Eisenstein series (10.3(1)) accordingly breaks into two partial sums E_1 and E_w. The first one reduces to $\varphi_{P,s}$, which is N-invariant on the left, so $E_{1,P} = \varphi_{P,s}$. In order to prove (i), we claim that it suffices to show:

(1) $$E_{w,P}(ax) = h_P(a)^{1-s}E_{w,P}(x) \quad (a \in A, \; x \in G).$$

Assume that (1) holds. Then, in view of 2.8(4), $E_{w,P}.h_P^{s-1}$ is left-invariant under A. On the other hand, it is N-invariant on the left and of K-type m on the right (see 10.2(8)). It therefore belongs to $F(P, m)$, so that we can write

$$E_{w,P} = c_{P|P}(s)\varphi_{P,-s} \quad (\sigma > 1)$$

for some well-defined function $c_{P|P}$ of s, which shows that $E_{w,P}$ is the second summand in 11.2(1). There remains to prove (1). We claim first that if $\gamma \in \Gamma_w$ and $\delta, \delta' \in \Gamma_N$ then the equality $\Gamma_P.\gamma.\delta = \Gamma_P.\gamma.\delta'$ implies $\delta = \delta'$. Let us write

(2) $\gamma = u_\gamma . a_\gamma . w . n_\gamma \quad (u_\gamma \in N \cup (-1)N, \ a_\gamma \in A, \ n_\gamma \in N).$

Then the previously assumed equality becomes

(3) $u_\gamma a_\gamma w n_\gamma \delta = u_\gamma a_\gamma w n_\gamma \delta'$

modulo Γ_P on the left, and the uniqueness in the decomposition PwN forces $n_\gamma \delta = n_\gamma \delta'$ and hence $\delta = \delta'$. We therefore have

$$E_w(ax) = \sum_{\gamma \in \Gamma_P \backslash \Gamma_w / \Gamma_N} \left(\sum_{\delta \in \Gamma_N} \varphi_{P,s}(\gamma \delta n a x) \right),$$

$$E_{w,P}(ax) = \sum_{\gamma \in \Gamma_P \backslash \Gamma_w / \Gamma_N} \sum_{\delta \in \Gamma_N} \int_{\Gamma_N \backslash N} \varphi_{P,s}(\gamma \delta n a x) \, dn,$$

(4) $$E_{w,P}(ax) = \sum_{\gamma \in \Gamma_P \backslash \Gamma_w / \Gamma_N} \int_N \varphi_{P,s}(\gamma n a x) \, dn.$$

To prove (1), it thus suffices to show

(5) $$\int_N \varphi_{P,s}(\gamma n a x) \, dn = h_P^{1-s}(a) \int_N \varphi_{P,s}(\gamma n x) \, dn \quad (\gamma \in \Gamma_w).$$

From (2) and 10.2 we derive

$$\varphi_{P,s}(\gamma n a x) = \varphi_{P,s}(u_\gamma . a_\gamma) \varphi_{P,s}(w . n_\gamma n . a . x).$$

However,

$$w . n_\gamma . n . a . x = w . a . {}^{a^{-1}}n_\gamma . {}^{a^{-1}}n . x.$$

In view of $w . a . w^{-1} = a^{-1}$, this can be written as

$$w . n_\gamma . n . a . x = a^{-1} w . {}^{a^{-1}}n_\gamma . {}^{a^{-1}}n . x,$$

whence

(6) $\varphi_{P,s}(w . n_\gamma . n . a . x) = h_P^{1+s}(a^{-1}) . \varphi_{P,s}(w . {}^{a^{-1}}n_\gamma . {}^{a^{-1}}n . x),$

(7) $\varphi_{P,s}(\gamma n a x) = h_P^{1+s}(a^{-1}) \varphi_{P,s}(u_\gamma . a_\gamma . w . {}^{a^{-1}}n_\gamma . {}^{a^{-1}}n . x),$

(8) $\varphi_{P,s}(\gamma n a x) = h_P^{1+s}(a^{-1}) \varphi_{P,s}(\gamma . n_\gamma^{-1} . {}^{a^{-1}}n_\gamma . {}^{a^{-1}}n . x).$

We now make the change of variables

$$n' = n_\gamma^{-1} . {}^{a^{-1}}n_\gamma . {}^{a^{-1}}n$$

on N. The translation by the two first factors on the right has no effect on the Haar measure. On the other hand, Int a is the dilation by $h_P(a)^2$ on N, so

(9) $dn = h_P(a)^2 . dn'$

and (5) follows from (8) by integration over N.

(iii) Using 10.4, we have

(10) $|E(P, s)_{P'}(x)| \leqslant \displaystyle\int_{\Gamma_{N'}\backslash N'} E_0(nx)\, dx$

$$\prec \frac{c^\sigma}{\sigma - 1} h_{P'}(x)^{\sigma+1} \quad (\sigma > 1,\ x \in \mathfrak{S}').$$

If $P' = P$ then, by (i),

(11) $|c_{P|P}(s) h_{P'}^{1-\sigma}(x)| \leqslant h_{P'}^{1+\sigma}(x) + \dfrac{c^\sigma}{\sigma - 1} h_{P'}(x)^{\sigma+1};$

if P' is not Γ-conjugate to P, then

(12) $|c_{P|P'}(s) h_{P'}^{1-\sigma}(x)| \leqslant \dfrac{c^\sigma}{\sigma - 1} h_{P'}(x)^{\sigma+1}.$

Fix $x \in \mathfrak{S}'$ such that $h_{P'}(x) > 1$. Since $c^\sigma(\sigma - 1)^{-1}$ has a strictly positive lower bound ($c > 1$), (iii) follows from (11) and (12).

In particular, $|c_{P|P'}(s)|$ is bounded on any vertical strip $\sigma \in [d, d']$ if

$$1 < d \leqslant d' < \infty.$$

11.4 A "continuation" principle. If U is a relatively compact domain in $\mathbb{C}$, we let $\mathbb{C}_\circ(U)$ denote the field of meromorphic functions on U that have finitely many zeroes and poles, with values in some topological vector space, as in 11.0. (We could as well define it as the field of meromorphic functions, each of which is the restriction of a meromorphic function on some domain containing the closure of U in its interior.) The principle may be stated as follows.

(∗) *Let F be a function holomorphic in a domain D of $\mathbb{C}$. Assume that F has been meromorphically continued to a meromorphic function F_1 on some domain $D_1 \supset D$. Let $s_\circ \notin D_1$ and assume we have found an open disc U around $s_\circ$ such that $U \cap D_1$ contains some disc U' and that U and U' have the following property: There exists a finite dimensional vector space V over $\mathbb{C}_\circ(U)$ and an infinite nonhomogeneous system $\mathcal{L}$ of linear equations in V, with coefficients in $\mathbb{C}_\circ(U)$ such that in U', the system $\mathcal{L}$ has a unique solution, which is the restriction to U' of F_1. Then $\mathcal{L}$ has a unique solution on U, which provides a meromorphic continuation of $F_1 \mid D \cap U$ to U.*

This follows from Cramer's rule: let d be the dimension of V over $\mathbb{C}_\circ(U)$. The assumption implies the existence of a subsystem $\mathcal{L}'$ of d equations in $\mathcal{L}$, some nonhomogeneous, which has a unique solution in U'. In particular, the determinant D_1 of the coefficients of the unknowns in those equations is not equal to zero in U' and hence is a meromorphic function on U. The expression of the solution as a quotient of a determinant by D_1, by Cramer's rule, provides the desired extension.

Our U will be a disc for which there exists a compatible $\alpha \in I_c^\infty(G)$ (10.7). We now make some preparations to define the system $\mathcal{L}$, the space of values for our meromorphic functions, and to prove a uniqueness assertion.

11.5   We must remember that the choice of a right K-type m underlies the construction of an Eisenstein series. As before, we omit the m in order to ease the notational burden. We let δ be the element of $\{0, 1\}$ which is $\equiv m \bmod 2$; the numbering of the cusps is as in 10.12. We write, respectively,

$$E_j, \; \varphi_{i,s}, \; c_{j,i}, \; d_{j,i}, \; h_i \quad \text{for} \quad E(P_j), \; \varphi_{P_i,s}, \; c_{P_j|P_i}, \; d_{P_j|P_i}, \; h_{P_i}.$$

By 11.2,

(1) $$E_j(s)_{P_i} = \delta_{j,i}\varphi_{i,s} + c_{j,i}(s).\varphi_{i,-s}.$$

By 10.12(1) and (2),

(2) $$E_j(s)_{P_i} = 0 \quad \text{if either } i > l(\delta) \text{ or } j > l(\delta)$$

This is of course an empty statement if m is even.

Let $\mathrm{E}(s)$ be the column vector with entries $E_1(s), \ldots, E_{l(\delta)}(s)$, and let $\mathrm{C}(s)$ be the matrix $(c_{i,j}(s))$ $(1 \leqslant i, j \leqslant l(\delta))$. (The size of the matrix depends therefore on the parity of m.) These are holomorphic for $\mathcal{R}s > 1$. Our goal is to show that E and C have meromorphic continuation to $\mathbb{C}$ and satisfy the functional equations

(3) $$\mathrm{E}(-s) = \mathrm{C}(-s)\mathrm{E}(s), \qquad \mathrm{C}(-s).\mathrm{C}(s) = 1.$$

We put ourselves in the situation of 9.6. Let $\pi: G \to \Gamma \backslash G$ be the canonical projection, and let $\mathfrak{S}_{i,t}$ be a Siegel set with respect to P_i, with t big enough so that the cusps $\mathfrak{S}_i = \pi(\mathfrak{S}_{i,t})$ are disjoint. We also assume that π induces a homeomorphism of $\Gamma_{N_i} \backslash N_i \mathfrak{S}_{i,t}$ onto $\mathfrak{S}_i$ (which is possible; see 3.16). Let moreover χ_i be the characteristic function of $\mathfrak{S}_i$ on $\Gamma \backslash G$. We let D be the complement of the union of the $\mathfrak{S}_i$, and set

$$H_D = \{f \in L^2(\Gamma \backslash G), \text{ of right } K\text{-type } m \text{ and } f_{P_i}|\mathfrak{S}_i = 0, \; i = 1, \ldots, l\};$$

H_D is the space that was denoted $°L^2(\Gamma \backslash G, D)$ with the conventions of 9.6. By 9.7, H_D is closed in $L^2(\Gamma \backslash G)$.

11.6 The truncation operator.   We let Λ be the map that assigns, to a locally L^1 function f on $\Gamma \backslash G$ of right K-type m, the function

(1) $$\Lambda f = f - \sum_i \chi_i f_{P_i},$$

where χ_i is the characteristic function of $\mathfrak{S}_{i,t}$.

This is a so-called truncation operator, denoted Λ^t, if the t underlying the definition of the Siegel sets $\mathfrak{S}_{i,t}$ (in 11.5), and hence of the χ_i, needs to be specified. Since $(f_P)_P = f_P$ the operator Λ is idempotent. Assume now that

$$f \in L^2(\Gamma \backslash G).$$

Then so is Λf and, by the remark just made, $\Lambda f \in H_D$. We claim that Λ is the orthogonal projection of $L^2(\Gamma \backslash G)$ onto H_D. Since Λ is idempotent, this amounts to showing that it is self-adjoint. Let $f, g \in L^2(\Gamma \backslash G)$. Then

$$(\Lambda f, g) = (f, g) - \sum_i (\chi_i f_{P_i}, g), \qquad (f, \Lambda g) = (f, g) - \sum_i (f, \chi_i g_{P_i}).$$

Therefore, it suffices to show that

(2) $$(\chi_i f_{P_i}, g) = (f, \chi_i g_{P_i}) \quad (i = 1, \dots, l).$$

This is a simple computation:

$$(\chi_i f_{P_i}, g)$$

$$= \int_{\Gamma \backslash G} \chi_i f_{P_i}(x) \bar{g}(x) \, dx = \int_{\mathfrak{S}_i} f_{P_i}(x) \bar{g}(x) \, dx$$

$$= \int_{\mathfrak{S}_i} \bar{g}(x) \, dx \int_{\Gamma_{N_i} \backslash N_i} f(nx) \, dn = \int_{\Gamma_{N_i} \backslash N_i} dn \int_{\mathfrak{S}_i} \bar{g}(x) f(nx) \, dx$$

$$= \int_{\Gamma_{N_i} \backslash N_i} dn \int_{\mathfrak{S}_i} \bar{g}(n^{-1}y) f(y) \, dy = \int_{\mathfrak{S}_i} f(y) \, dy \left(\int_{\Gamma_{N_i} \backslash N_i} \bar{g}(n^{-1}y) \, dn \right)$$

$$= \int_{\mathfrak{S}_i} f(y) \bar{g}_{P_i}(y) \, dy = (f, \chi_i g_{P_i}).$$

Thus, Λ is the operator denoted Q in 9.8, to which we can therefore apply 9.8. Hence, if $\alpha \in I_c^\infty(G)$ then

(3) $$*\alpha \colon H_D \to L^2(\Gamma \backslash G) \quad \text{and} \quad (*\alpha) \circ \Lambda \colon L^2(\Gamma \backslash G) \to L^2(\Gamma \backslash G)$$

are compact operators. Similarly,

(4) $$\Lambda \circ (*\alpha) \colon H_D \to H_D$$

is a compact operator, since it is the composition of a compact one and a continuous one.

Lemma. *Let f be a locally L^1 function on $\Gamma \backslash G$, of right K-type m and moderate growth. Let $\alpha \in I_c^\infty(G)$. Then $\Lambda(f * \alpha) \in H_D$.*

The difference $f * \alpha - (f * \alpha)_{P_i}$ is rapidly decreasing on $\mathfrak{S}_i$ (7.6) and hence, a fortiori, square integrable on $\mathfrak{S}_i$ $(i = 1, \dots, l)$. Because $f * \alpha$ is smooth, it is square integrable on D. This proves the lemma.

11.7   We need one more fact about compact operators. Let again H be a separable Hilbert space, $(\ ,\ )$ the scalar product on H, and $\mathcal{L}(H)$ the algebra of bounded operators on H. The latter has a natural structure of Banach space with respect to the norm topology [46, VI.1]. Let A be a linear operator on H (perhaps unbounded but densely defined). The resolvent set $\rho(A)$ of A is the set of $\lambda \in \mathbb{C}$ for which $(\lambda I - A)$ has a dense range and a bounded inverse, denoted $R(\lambda, A)$. The complement of $\rho(A)$ is the *spectrum* sp(A) of A. If $\ker(\lambda I - A) \neq 0$, then λ is said to belong to the *discrete spectrum* $\sigma(A)$, and a nonzero element of that kernel is an eigenvector of A for λ. Clearly $\sigma(A) \subset$ sp(A). Assume that A is closed (in particular, continuous); then $\rho(A)$ is open and $s \mapsto R(s, A)$ is a holomorphic function on $\rho(A)$, with values in the Banach space $\mathcal{L}(H)$ [61, VIII.2, p. 211, Thm. 1].

   Assume now that A is compact. Then sp$(A) \subset \sigma(A) \cup \{0\}$ [61, X, §5, p. 283, Thm. 1]; hence $s \mapsto R(s, A)$ is holomorphic outside $\sigma(A) \cup \{0\}$. If, moreover, A is self-adjoint, then $\sigma(A) \subset \mathbb{R}$ and hence $R(s, A)$ is holomorphic in $\mathbb{C}^* - \mathbb{R}^*$.

11.8 The main construction.   We choose an open disc $U \subset \mathbb{C}$ and $\alpha \in I_c^\infty(G)$ compatible with one another (10.7). Note that $\pi_\alpha(s) \neq 0$ on U and that it takes a given value only finitely many times. Therefore, 11.7 also implies that

$$(1) \qquad\qquad s \mapsto (\Lambda \circ (*\alpha) - \pi_\alpha(s))^{-1} \quad (s \in U).$$

is a meromorphic family of bounded operators on H_D.

   Given a set $\mu = (\mu_{\pm 1}, \dots, \mu_{\pm l(\delta)})$ of $2l(\delta)$ elements of $\mathbb{C}_o(U)$, we let

$$(2) \qquad\qquad \Psi_\mu(s) = \sum_{1 \leqslant i \leqslant l(\delta)} (\mu_i \chi_i \varphi_{i,s} + \mu_{-i} \chi_i \varphi_{i,-s}).$$

For fixed s, this is a moderately increasing function on $\Gamma \backslash G$.

A. Lemma.   *Given* $\mu = (\mu_{\pm i}) \in \mathbb{C}_o(U)^{2l(\delta)}$, *there exists a unique meromorphic function* $F_\mu(s)$ *from* U *to moderately increasing functions on* $\Gamma \backslash G$ *of right K-type m that satisfies the conditions*

(i) $s \mapsto F_\mu(s) - \Psi_\mu(s)$ *is a meromorphic map from* U *to* H_D,
(ii)$_\alpha$ $\Lambda(F_\mu(s) * \alpha) = \pi_\alpha(s)\Lambda F_\mu(s)$.

Proof. We note first that, since $\varphi_{\pm s}$ is N-invariant on the left, it is equal to its constant term; therefore,

$$(3) \qquad\qquad\qquad \Lambda(\Psi_\mu(s)) = 0.$$

Assume there are functions F_s and $g_s \in H_D$ such that

$$F_s = g_s + \Psi_\mu(s)$$

satisfies (ii)$_\alpha$. Then, using (3), we obtain

(4) $\qquad \Lambda(F_s * \alpha) = \Lambda(g_s * \alpha) + \Lambda(\Psi_\mu(s) * \alpha) = \pi_\alpha(s)\Lambda g_s = \pi_\alpha(s)g_s$

(recall that Λ is the identity on H_D), which can be written

(5) $\qquad\qquad (\Lambda \circ (*\alpha) - \pi_\alpha(s))g_s = -\Lambda(\Psi_\mu(s) * \alpha).$

By the lemma in 11.6, $\Lambda(\Psi_\mu(s) * \alpha) \in H_D$. Therefore (cf. (1)), we can write

(6) $\qquad\qquad g_s = -(\Lambda \circ (*\alpha) - \pi_\alpha(s))^{-1}.\Lambda(\Psi_\mu(s) * \alpha).$

Hence there is at most one $g_s \in H_D$, so (i) and (ii) are satisfied. Conversely, define g_s by (6). Then (5) holds and, in view of (3),

$$F_s = g_s + \Psi_\mu(s)$$

satisfies (ii)$_\alpha$. By construction, (i) holds.

B. Linear equations. Note that g_s, as defined by (6), is linear in the $\mu_{\pm i}$ and of right K-type m; that is, we can write it as a sum

(7) $\qquad\qquad\qquad g_s = \sum_{\pm i} v_{\pm i}(s)\mu_{\pm i}(s),$

where $s \mapsto v_{\pm i}(s)$ is a meromorphic map in H_D. Here $1 \leqslant i \leqslant l(\delta)$.

We now impose on $F = F_\mu(s)$ the condition

(8) $\qquad\qquad\qquad (\mathcal{C} - (s^2 - 1)/2)F_\mu(s) = 0.$

Since F_s is not smooth on $\Gamma \backslash G$, this must be viewed in the distribution sense, that is,

(9)$_\varphi$ $\qquad \displaystyle\int_G F_\mu(s, x).(\mathcal{C} - (s^2 - 1)/2)\varphi(x)\,dx = 0 \quad (\varphi \in C_c^\infty(G)).$

Taking (7) into account, we see that (9)$_\varphi$ is a homogeneous linear equation in the $\mu_{\pm i}$, with coefficients in $C_\circ(U)$. For $j = 1, \ldots, l(\delta)$, we let $\mathcal{L}_j$ be the system of linear equations in the $\mu_{\pm i}$ consisting of all equations (9)$_\varphi$ ($\varphi \in C_c^\infty(G)$), of the conditions (i), and finally of

(10) $\qquad\qquad\qquad \mu_i = \delta_{ij} \quad (i = 1, \ldots, l(\delta)).$

Note that (i) for F_s and (10) for μ imply that the constant term of F_s on the cusp $\mathfrak{S}_i$ is the sum of $\delta_{i,j}\varphi_{i,s}$ and of the product of $\varphi_{i,-s}$ by a meromorphic function μ_{-i} in s.

In the sequel, we let $\mathcal{L}_{j,\alpha}$ be the set of conditions on $F(s)$ and its constant term consisting of (i), (ii)$_\alpha$, and $\mathcal{L}_j$.

C. Lemma. *Assume that U is contained in the half-plane $\mathcal{R}s > 1$. Then $E_j(s)$ is the only function satisfying $\mathcal{L}_{j,\alpha}$ for a given $\alpha \in I_c^\infty(G)$ compatible with U. It satisfies $\mathcal{L}_{j,\beta}$ for any $\beta \in I_c^\infty(G)$ compatible with U, and is holomorphic with values in $C^\infty(G)$ (see 11.0).*

Proof. The Eisenstein series $E_j(s)$ is a smooth eigenfunction of $\mathcal{C}$ with eigenvalue $(s^2 - 1)/2$. By 11.3, it is of the form $F_s(\mu)$, with μ satisfying (10); it therefore satisfies $\mathcal{L}_j$. By 10.5,

$$(11) \qquad\qquad E_j(s) * \beta = \pi_\beta(s) E_j(s)$$

for any $\beta \in I_c^\infty(G)$. This implies (ii)$_\beta$. Therefore, $E_j(s)$ is a solution on U of all the systems $\mathcal{L}_{j,\beta}$. Now fix $\alpha \in I_c^\infty(G)$ compatible with U, and assume that $F(s)$ also satisfies $\mathcal{L}_{j,\alpha}$. Note that since $F(s)$ is K-finite, $\mathcal{Z}$-finite, and of moderate growth, it is an automorphic form. Let $R(s) = E_j(s) - F(s)$. Then, for fixed $s \in U$, $R(s)_{P_i}$ is a constant multiple of $\varphi_{i,-s}$ (recall that $E_j(s)_{P_i}$ and $F(s)_{P_i}$ both satisfy (10)). Since $\mathcal{R}s > 1$, $\varphi_{i,-s}$ is square integrable on $\mathfrak{S}_i$. But then so is $R(s)$, since $R(s) - R(s)_{P_i}$ $(i = 1, \ldots l)$ is rapidly decreasing (7.8). Therefore $R(s)$ is square integrable; it is an eigenfunction of $\mathcal{C}$ with eigenvalue $(s^2 - 1)/2$. This forces s^2 to be real (see 11.12) and hence also s to be real (it cannot be purely imaginary since $\sigma > 1$) unless $R(s) = 0$. So $R(s) = 0$: first for $s \in U$, $s \notin \mathbb{R}$; and then identically in U.

As we saw in the proof of 10.4, for s and x in bounded regions of $\mathcal{R}s > 1$ and of G, respectively, the series defining $E_j(s)$ has a constant termwise absolute majorant. Since h_P^{1+s} is entire in s, the last assertion follows.

11.9 Theorem. *Let E and C be as in 11.5. Then E and C admit meromorphic continuations to $\mathbb{C}$ that satisfy the functional equations*

$$(*) \qquad\qquad \mathsf{E}(-s) = \mathsf{C}(-s).\mathsf{E}(s), \qquad \mathsf{C}(-s).\mathsf{C}(s) = 1.$$

If $E_j(s)$ has a pole of order $m(c)$ at $c \in \mathbb{C}$, then $(s - c)^{m(c)}.E_j(s)$ is, at c, an automorphic form of right K-type m and an eigenfunction for $\mathcal{C}$ with eigenvalue $(c^2 - 1)/2$ $(1 \leqslant j \leqslant l(\delta))$.

Proof. We assume first that $s, c \neq 0$. Let V be a domain in $\mathbb{C}$ containing the half-plane $\sigma > 1$. We assume that E_j has been meromorphically continued to V (in the sense of 11.0); as a result, the constant term E_{j,P_i} $(i = 1, \ldots, l(\delta))$ has also been so continued. For $s \neq 0$, it can be written uniquely as $d_{i,j}(s)\varphi_{i,s} + c_{j,i}(s)\varphi_{i,-s}$. Since $\varphi_{i,\pm s}$ is entire and nowhere zero, this determines uniquely the coefficients, which are then analytic continuations of those coefficients given for $\sigma > 1$. In particular $d_{j,i}(s) = \delta_{j,i}$. We also assume that, in a neighborhood of any point of V, all the conditions satisfied by E_j in $\sigma > 1$ (see 11.8(C)) also hold. Let

$$\Psi_j(s) = \sum_i (\delta_{i,j}\varphi_{i,s} + c_{i,j}(s)\varphi_{i,-s})\chi_i.$$

This is our previous $\Psi_\mu(s)$, with

$$\mu_i = \delta_{i,j}, \mu_{-i} = c_{i,j}(s).$$

Then the map $s \mapsto (E_j(s) - \Psi_j(s)) \in H_D$ is meromorphic.

Let now c be on the boundary of V, and let U be a disc centered on c such that $V \cap U$ contains a disc U' for which there is a compatible $\alpha \in I_c^\infty(G)$. By choosing U' suitably we may assume that $E_j(s)$ is holomorphic in U'. The function α is also compatible with U' so that, by assumption, $E_j(s)$ is the only solution of $\mathcal{L}_{j,\alpha}$ in U'. By the continuation principle 11.4, this solution has a meromorphic continuation, which is also the unique solution of $\mathcal{L}_{j,\alpha}$ on U.

The relation 11.8(10), which is valid in $U \cap V$, remains valid in U. Let $c \in U$. Let $F_j(s) = E_j(s) - \Psi_j(s)$. By assumption there is a small disc $D(c, R)$ around c, $m(c) \in \mathbb{N}$ and elements $h_n \in H_D$ $(n \in \mathbb{N})$ such that

$$(1) \qquad (s - c)^{m(c)} F_j(s) = \sum_{n \geqslant 0} (s - c)^n . h_n$$

with

$$(2) \qquad \sum_{n \geqslant 0} R^n \|h_n\| < \infty.$$

Given $F \in \mathcal{U}(\mathfrak{g})$, by 9.7 there is a constant $C(\alpha, F) > 0$ such that

$$(3) \qquad |F(h_n * \alpha)(x)| \leqslant C(\alpha, F)\|h_n\| \quad (n \in \mathbb{N}, x \in \Gamma\backslash G),$$

so the series

$$\sum_{n \geqslant 0} (s - c)^n F(h_n * \alpha)$$

has a termwise absolutely convergent majorant and hence is absolutely and uniformly convergent on $D(c, R) \times (\Gamma\backslash G)$. As a consequence, $(s - c)^{m(c)} F_j * \alpha$ is a bounded holomorphic function from $D(c, R)$ to $C^\infty(\Gamma\backslash G)$. Clearly

$$(s - c)^{m(c)}.\Psi_j(s) * \alpha$$

is also holomorphic. But we have

$$(4) \qquad (s - c)^{m(c)} E_j(s) * \alpha = (s - c)^{m(c)}.\pi_\alpha(s)E_j(s)$$

(see 10.5), so that finally

$$(5) \qquad (s - c)^{m(c)} E_j(s) = \pi_\alpha(s)^{-1}(s - c)^{m(c)}(F_j(s) * \alpha + \Psi_j(s) * \alpha),$$

showing that the left-hand side is a holomorphic function on $D(c, R)$ with values in $C^\infty(G)$. This provides the meromorphic continuation to U.

Given $c \in \mathbb{C}$, we can find finitely many compatible pairs (U_i, α_i) $(i = 1, \ldots, m)$ such that U_1 is contained in $\sigma > 1$, U_m contains c, and

$$U_i \cap U_{i-1} \neq \emptyset \quad (i = 2, \ldots, m).$$

For $c \neq 0$, the foregoing yields the meromorphic continuation around c.

There remains the origin – but this is just a technicality. This limitation has arisen because of the way the constant term was written. For $s \neq 0$ the functions h_P^{1+s} and h_P^{1-s} are, on A, two (independent) solutions of 11.1(2). For $s = 0$ they coincide, the characteristic equation has a double solution, and the solutions of 11.1(2) are h_P and $h_P \log h_P$. It is easy to take care of this by writing the constant term in a way that remains valid at the origin.

Let

(6) $$\beta_{P,s} = (\varphi_{P,s} + \varphi_{P,-s})/2, \qquad \gamma_{P,s} = (\varphi_{P,s} - \varphi_{P,-s})/2s,$$

which can also be written

$$\beta_{P,s} = \varphi_{P,0}.h_P \cosh h_P^s, \qquad \gamma_{P,s} = \varphi_{P,0}.h_P \sinh h_P^s.s^{-1}.$$

These are also entire functions that for each s, including the origin, are linearly independent. We have

(7) $$\beta_{P,0} = \varphi_{P,0}, \qquad \gamma_{P,0} = \varphi_P.h_P \log h_P = \varphi_{P,0} \log h_P.$$

In 11.1(1), (4) the constant term of f can also be written as

(8) $$f_P = a.\beta_{P,s} + b.\gamma_{P,s}$$

$(a, b \in \mathbb{C})$, an expression which is valid for any $s \in \mathbb{C}$. We now come back to E_j, and write $\beta_{i,s}$ for $\beta_{P_i,s}$ and $\gamma_{i,s}$ for $\gamma_{P_i,s}$. The constant term takes the form

(9) $$E_j(s)_{P_i} = e_{j,i}(s).\beta_{i,s} + f_{j,i}(s)\gamma_{i,s} \quad (i = 1, \ldots, l(\delta)).$$

The relations 11.8(10) are replaced by

(10) $$s.e_{j,i}(s) + f_{j,i}(s) = 2s\delta_{j,i} \quad (i = 1, \ldots, l(\delta)),$$

which, again for $i = j$, yield a nonhomogeneous condition.

Choose a compatible pair (U, α), where U is a disc centered at the origin. For $\mu = (\mu_{\pm i}) \in C_o(U)^{2l(\delta)}$, we now define

(11) $$\tilde{\Psi}_j(s) = \sum_1^l (\mu_i \chi_i \beta_{i,s} + \mu_{-i} \chi_i \gamma_{i,s})$$

and let $\tilde{\mathcal{L}}_j$, $\tilde{\mathcal{L}}_{j,\alpha}$ denote the same set of conditions as before, except that 11.8(10) is replaced by (10). In any disc U' in the interior of U, not containing the origin,

$\tilde{\mathcal{L}}_j$ and $\tilde{\mathcal{L}}_{j,\alpha}$ are obviously equivalent to $\mathcal{L}_j$ and $\mathcal{L}_{j,\alpha}$. The meromorphic continuation of E_j in U', the existence of which has already been proved, is then also the unique solution of $\tilde{\mathcal{L}}_{j,\alpha}$. The principle of 11.4 provides a unique meromorphic extension to U. At first, it involves a meromorphic function with values in H_D, but our previous argument – showing that we obtain a meromorphic function with values in $C^\infty(G)$ – applies verbatim here.

We now prove that the determinant det C of the matrix C is not identically zero. Assume it is. Then we can find meromorphic functions m_j on $\mathbb{C}$ $(j = 1, \ldots, l(\delta))$, not all zero, such that

$$\sum_j m_j . c_{j,i} = 0 \quad (i = 1, \ldots, l(\delta)).$$

Let $R = \sum_j m_j E_j$. Then

$$R(s)_{P_i} = \sum_j m_j(s) E_j(s)_{P_i} = \sum_j m_j(s)(\delta_{j,i}\varphi_{i,s} + c_{j,i}(s)\varphi_{i,-s}),$$

(12) $$R(s)_{P_i} = m_i(s)\varphi_{i,s} \quad (\text{on } \mathfrak{S}_i).$$

Let U be a domain in the left half-plane $\sigma < 0$ on which the m_j and E_j are holomorphic in s. For $s \in U$, the function $\varphi_{i,s}$ tends to zero on $\mathfrak{S}_i$; hence so does $R(s)$ (9.3) and, as a consequence, $R(s)$ is square integrable. But the $E_j(s)$ are eigenfunctions of $\mathcal{C}$ with eigenvalue $(s^2 - 1)/2$, hence so is $R(s)$ $(s \in U)$. Therefore s should be real (11.7) unless $R(s) = 0$. Consequently $R(s)$ is identically zero; that is, we have a relation

(13) $$\sum_j m_j(s) E_j(s) = 0$$

for all $s \in \mathbb{C}$ at which the m_j and E_j are holomorphic. However, (12) and (13) imply that $m_j(s) = 0$ $(j = 1, \ldots, l(\delta))$, a contradiction with our initial assumption. Therefore, det C is not identically zero and we can form the inverse matrix $C(s)^{-1}$. Its coefficients are also meromorphic functions. It is now easy to prove the functional equation as follows.

Consider the column vector $Q(s) = C(-s)^{-1}.E(-s)$ with entries $Q_i(s)$. We have

$$Q_i(s) = \sum_m C^{-1}(-s)_{i,m} E_m(-s);$$

therefore,

$$Q_i(s)_{P_j} = \sum_m C^{-1}(-s)_{i,m}(\delta_{m,j}\varphi_{j,-s} + c_{m,j}(-s)\varphi_{j,s})$$

$$= C^{-1}(-s)_{i,j}\varphi_{j,-s} + \delta_{i,j}\varphi_{j,s}.$$

It follows that, in the neighborhood U of any point in which C, C^{-1}, and E are holomorphic, $Q_i(s)$ satisfies the conditions $\mathcal{L}_{j,\alpha}$ of 11.8 for all α compatible with U. Therefore, by what has been proved,

$$Q_i(s) = E_i(s) \quad (i = 1, \dots, l);$$

that is, $\mathsf{E}(s) = \mathsf{Q}(s)$. The first relation in $(*)$ then follows from the definition of $\mathsf{Q}(s)$. Moreover, $E_i(s)$ and $Q_i(s)$ have the same constant terms, so

$$c_{i,j}(s) = \mathsf{C}^{-1}(-s)_{i,j},$$

which yields the second part of $(*)$.

We still need to prove the last assertion. It is clear that $(s - c)^{m(c)}.E_j(s)$ is of right K-type m and is an eigenfunction for $\mathcal{C}$ with eigenvalue $(s^2 - 1)/2$ for every $s \in D(c, R)$. There remains to show that it has moderate growth. In fact, we want to prove more strongly the existence of strictly positive constants C, N such that

$$(14) \qquad |(s - c)^{m(c)} E_j(s)(x)| \leqslant C \|x\|^N \quad (s \in D(c, R),\ x \in \Gamma \backslash G).$$

Consider again (5). Since $(s - c)^{m(c)} E_j(s)$ is holomorphic for $s \in D(c, R)$, its constant terms at the P_i are linear combinations with bounded coefficients of the functions $\varphi_{P,s}$, $\varphi_{P,-s}$, and it is clear that, on a Siegel set, those have growth that is bounded by a function $h_P^d(g)$ for $d > \max(1 + \sigma,\ 1 - \sigma)$ $(s \in D(c, R))$. The same is then true for the $(s - c)^{m(c)} E_{j,P_i}$. Next, we have seen that $|F_j(s)(x)|$ is bounded for $s \in D(c, R)$ and $x \in \Gamma \backslash G$. Thus (14) is satisfied on a Siegel set with respect to P_i for all i and hence on $\Gamma \backslash G$. Since $|\pi_\alpha(s)|^{-1} \leqslant 2$ on $D(c, R)$ by construction, the last assertion now follows from (5), with a rate of growth bounded for all $s \in D(c, R)$. Therefore, this also implies the following.

11.10 Corollary. *Let U be a relatively compact open subset of $\mathbb{C}$, and let $c \in U$. Assume that $E_j(s)$ is holomorphic at all points of the closure of U except for a pole of order $m(c)$ at c. Let $\mu \in C_c(\Gamma \backslash G)$. Then the function $s \mapsto (s - c)^{m(c)} \int_{\Gamma \backslash G} \mu(g) E_j(s)(g)\, dg$ is holomorphic on U.*

11.11 Theorem. *Let $c \in \mathbb{C}^*$ and $n \in \mathbb{N}$. Let $E(s) = \sum_j \mu_j E_j(s)$.*

  (i) *Assume that $E(s)$ has a pole of order n at c. Then the automorphic form $(s - c)^m.E(s)$ for $s = c$ is orthogonal to the cusp forms.*
 (ii) *$E(s)$ has a pole of order $\leqslant n$ at c if and only if all its constant terms $E(s)_{P_i}$ have poles of orders $\leqslant n$ at c.*

Proof. (i) We can find a domain $C \subset \mathbb{C}$ that contains a non-empty open subset of the half plane $\mathcal{R}s > 1$ and in which all the $E_j(s)$ are holomorphic, except possibly for the pole at c. Let f be a cusp form. Consider the function

$$s \mapsto ((s-c)^n.E(s), f)_{\Gamma \backslash G},$$

which is a holomorphic function on C. By 10.11, it is equal to zero if $s \in C$ and $\mathcal{R}s > 1$; it is therefore identically zero.

(ii) The definition of the constant term shows that, if E has a pole of order $\leqslant n$ at c, then so do its constant terms. Assume now that the constant terms have poles of orders $\leqslant n$ at c, and that $E(s)$ has a pole of order $q > n$ at c. Then, for $s = c$, the function $(s - c)^q.E(s)$ is an automorphic form all of whose constant terms are zero, that is, a cusp form. By (i) this form is orthogonal to itself and hence is zero – a contradiction.

Remark. Let $E(s) = E_j(s)$. The constant term $E_j(s)_{P_i}$ has a pole of order n at c if and only if $c_{j,i}(s)$ has a pole of order n at c. Therefore (ii) can also be expressed by saying that $E_j(s)$ has a pole of order $\leqslant n$ at c if and only if the $c_{j,i}(s)$ have poles of orders $\leqslant n$ at c $(i = 1, \ldots, l(\delta))$.

11.12   There remains to justify one assertion made in 11.8(C) – namely, that if a nonzero square integrable automorphic form φ is an eigenfunction for C with eigenvalue λ, then λ is real.

Let us write H for $L^2(\Gamma \backslash G)$, $(,)$ for the scalar product

$$(\varphi, \psi) = \int_{\Gamma \backslash G} \varphi(g)\overline{\psi(g)} \, dg,$$

and H^∞ for the space of elements in H represented by smooth functions φ, which – together with all their derivatives $D\varphi$ $(D \in \mathcal{U}(\mathfrak{g}))$ – are square integrable. This space contains all elements

(1) $f * \alpha$ $(f \in H, \alpha \in C_c^\infty(G))$.

(Actually, it follows from [16] that it is spanned by those, but we shall not need this fact.) By 5.6, this space contains all square integrable automorphic forms. We claim first that

(2) $(X\varphi, \psi) = -(\varphi, X\psi)$ $(X \in \mathfrak{g}, \varphi, \psi \in H^\infty)$.

This is a simple computation based on the fact that the right translations r_g $(g \in G)$ define unitary operators on H. Indeed, let r_t be the right translation by e^{tX} $(t \in \mathbb{R})$. Then $(r_t(\varphi), \psi) = (\varphi, r_{-t}(\psi))$, whence

(3) $(r_t(\varphi) - \varphi, \psi) = (\varphi, r_{-t}(\psi) - \psi)$;

(2) then follows from (3) and 2.1(6).

As a consequence,

$$(X.Y\varphi, \psi) = (\varphi, Y.X\psi)   (X, Y \in \mathfrak{g}, \varphi, \psi \in H^\infty).$$

But $C = H^2/2 + EF + FE$ (see 2.5), whence

(4) $(C\varphi, \psi) = (\varphi, C\psi).$

Therefore, if $C\varphi = \lambda\varphi$ ($\varphi \neq 0$) then λ is real.

We shall return to (4) in 13.3, in the discussion of the essential self-adjointness of C.

So far, no information is given on the nature of the poles of $C(s)$ or $E(s)$. However, some general facts are known in the right half-plane $\mathcal{R}s \geqslant 0$, which may be summarized as follows.

11.13 Theorem. (i) *We have* ${}^t\overline{C(\bar{s})} = C(s)$ *whenever* $C(s)$ *is regular at* s *and* $\bar{s}$. *The matrix* $C(s)$ *is unitary and holomorphic on the imaginary axis.*

(ii) *The functions* $E(P_i, s)$ *and* $c_{i,j}(s)$ *are holomorphic in* $\mathcal{R}s \geqslant 0$, *except for (at most) finitely many simple poles in* $(0, 1]$. *At such a pole, the residue of* $E(P_i, s)$ *is a square integrable automorphic form.*

(iii) *Let* $\sigma, T > 0$. *Then* $|c_{i,j}(s)|$ *is bounded in the region* $\mathcal{R}s \in [0, \sigma]$ *and* $|\mathcal{I}s| \geqslant T$.

The proof requires some further preparation and will be given in Section 12 (see 12.7 for (i) and 12.11 for (ii), and (iii)).

11.14 Proposition. *Let* $E(s) = \sum_i \mu_i.E_i(s)$ ($\mu_i \in \mathbb{C}$, $i = 1, \ldots, l(\delta)$) *and* $z \in \mathbb{C}^*$. *Assume that* $E(s)$ *is holomorphic at* z.

(i) *If the automorphic form* $E(z)$ *is zero, then* $E(s) = 0$ *for all* $s \in \mathbb{C}$.
(ii) *Assume the* $E_i(s)$ *are holomorphic at* z. *Then they are linearly independent over* $\mathbb{C}$.

Proof. (i) Let s run through a neighborhood of z in $\mathbb{C}^*$ in which the $E_i(s)$ are holomorphic except possibly for a pole at z. Fix $j \in [1, l(\delta)]$. We have

$$E(s)_{P_j} = \mu_j.\varphi_{P_{j,m,s}} + \sum_i \mu_i.c_{i,j}(s)\varphi_{P_{j,m,-s}}.$$

(We recall that m is the right K-type of the E_i, which has been fixed once for all.) Then

$$E(s)_{P_j} = \mu_j\varphi_{P_j,m,s} + d_j(s)\varphi_{P_{j,m,-s}} = (\mu_j.h_{P_j}^{1+s} + d_j(s)h_{P_j}^{1-s})\varphi_{P_{j,m}}$$

where $d_j(s) = \sum \mu_i c_{i,j}(s)$ is holomorphic around z.

Assume now that $E(z) = 0$. Then $E(z)_{P_i} = 0$, whence, on A_j,

$$\mu_j.h_{P_j}^{1+z} + d_j(z)h_P^{1-z} = 0.$$

Since $z \neq 0$, this forces $\mu_j = 0$ and so proves (i).

The assertion (ii) is an obvious consequence.

Remark. At the origin, the $E_i(s)$ ($i \leqslant l(\delta)$) are holomorphic, as we shall see, but they are not necessarily linearly independent.

11.15 A growth estimate. We draw here a consequence of the proof of 11.9. It will play an essential one in the second part Section 12 and an accessory one in Section 17.

As before, $\Gamma \backslash G$ is the union of a compact set C and of disjoint Siegel sets $\mathfrak{S}_{i,t}$ at the cusps; χ_i is the characteristic function of $\mathfrak{S}_{i,t}$. Fix j. Let B be a compact subset of $\mathbb{C}$. Let $\{c_q\}$ ($q \in Q$) be the (finitely many) poles of $E_j(s)$ on B, let $m(c_q)$ be the order of the pole of E_j at c_q, and set

$$M(s) = \prod_{q \in Q} (s - c_q)^{m(c_q)}.$$

Then $M.E_j$ is holomorphic on B.

11.16 Proposition. *Let $d = \max_{s \in B} |\mathcal{R}s|$ and let t be as before. Then there exists a constant $q > 0$ such that the following conditions hold.*

(1) $$|M(s).E_j(s)(x)| \leqslant q \quad (s \in B, \, x \in C).$$

(2) $|M(s).E_j(s)(x)| \leqslant q h_i(x)^{1+d} \quad (s \in B, \, s \neq 0, \, x \in \mathfrak{S}_{i,t}, \, i = 1, \ldots l(\delta)).$

(3) *Given $N \in \mathbb{N}$, there exists a constant q_N such that*

$$M(s)|E_j(s)^t(x)| \leqslant q_N h_{P_i}(x)^{-N} \quad (s \in B, \, x \in \mathfrak{S}_{i,t}, \, i = 1, \ldots, l(\delta)).$$

(4) *If $0 \in B$ then, for any $d' > d$,*

$$|M(s)E_j(s)(x)| \leqslant q.h_i(x)^{1+d'} \quad (s \in B, \, x \in \mathfrak{S}_{i,t}, \, i = 1, \ldots, l(\delta)).$$

The existence of such an estimate for a given s follows from Section 7. The point of the proposition is that q, q_N, and d' are independent of $s \in B$ ($s \neq 0$ in (2)).

Proof. A neighborhood of B in which E_j has no new pole can be covered by finitely many discs U_ι ($\iota \in I$) that admit compatible α_ι, on the closure of which $M.E_j$ is still holomorphic. Moreover, we can arrange that U_ι contain at most one c_q and that, if it does, c_q is the origin of U_ι. To establish (1) and (2) it suffices to prove them for $s \in U_\iota$, with d replaced by the lower upper bound of $|\mathcal{R}s|$ on $B \cap U_\iota$. This brings us to the situation of the proof of 11.9, and we go back to the notation there. Write $D(c, R)$ for U_ι, assume E_j to be holomorphic on $D(c, R)$ except possibly at the origin, and let $m(c)$ be the order of its pole (if any) at the origin and be equal to zero otherwise. Then, to establish (1) and (2), it suffices to prove the existence of a constant q such that

(5) $\qquad |(s - c)^{m(c)} E_j(s)(x)| \leqslant q \quad (s \in B \cap D(c, R), \, x \in C)$

and

(6)
$$|(s - c)^{m(c)} E_j(x)| \leqslant q h_i(x)^{1+d}$$
$$(s \in B \cap D(c, R), \, s \neq 0, \, x \in \mathfrak{S}_{i,t}, \, i = 1, \dots, l),$$

where d is the lower upper bound of $\mathcal{R}s$ on $B \cap D(c, R)$.

As in 11.9, we write

(7) $\qquad (s - c)^{m(c)} E_j * \alpha = (s - c)^{m(c)} \Psi_j * \alpha + F_j(s) * \alpha,$

where $\Psi_j(s)$ is still a sum of constant terms on the $\mathfrak{S}_{i,t}$ and F_j is the truncation of E_j. The function $F_j * \alpha$ belongs to H_D. By 9.7, for any $R \in \mathcal{U}(\mathfrak{g})$, the function $(s - c)^{m(q)} . R(F_j(s) * \alpha)$ is bounded on $D(c, R) \times G$ and holomorphic in s. Its constant term on each $\mathfrak{S}_{i,t}$ is zero; therefore, for any $s \in D(c, R)$, it is fast decreasing. We claim that the rate of decay of $(s - c)^{m(c)} . (F_j(s) * \alpha)$ has an upper bound independent of $s \in D(c, R)$. In other words, given $N \in \mathbb{N}$ there exists a constant δ_N such that

(8)
$$|(s - c)^{m(c)} . (F_j(s) * \alpha)(x)| \leqslant q_N . h_i(x)^{-N}$$
$$(s \in D(c, R), \, x \in \mathfrak{S}_{i,t}, \, i = 1, \dots, l(\delta)).$$

To see this, we return to 7.4 and apply it to $f = (s - c)^{m(c)} . F_j(s)$. By our previous remark, the coefficient of $a(x)^{-\alpha}$ (which we now write $h_P(x)^{-1}$; see 2.8) is bounded uniformly on $D(c, R) \times G$. It follows that the implied constants in 7.5(1)–(4) are independent of $s \in D(c, R)$ and $x \in G$. Because the constant term is zero on $\mathfrak{S}_{i,t}$, our assertion follows from 7.5. In view of the definition of the truncation $E(s)'$, this proves (3) and shows that the growth of $(s - c)^{m(c)} . E_j(s) * \alpha$ is the same as that of $(s - c)^{m(c)} . \Psi_j(s) * \alpha$, which, in turn, is dominated by that of $(s - c)^{m(c)} . \Psi_j(s)$, up to a constant factor depending only on α. On $\mathfrak{S}_{i,t}$ we have

(9) $\quad (s - c)^{m(c)} . \Psi_j(s) = (s - c)^{m(c)} \varphi_{i,m}(\delta_{j,i} h_i^{1+s} + c_{j,i}(s) h_i^{1-s}) \quad (s \neq 0),$

the growth of which is dominated, up to a constant factor, by that of $h_i^{1+|\mathcal{R}s|}$. Of course, $\Psi_j(s)$ is bounded on $D(c, R) \times C$. This therefore establishes (5) and (6) with the left-hand sides replaced by $(s - c)^{m(c)} . E_j(s) * \alpha$. However (11.9(4)),

(10) $\qquad (s - c)^{m(c)} E_j(s) * \alpha = (s - c)^{m(c)} . \pi_\alpha(s) . E_j(s),$

where $|\pi_\alpha(s)|$ is bounded away from zero on $D(c, R)$, since α is assumed to be compatible with $D(c, R)$; hence (5) and (6) for $(s - c)^{m(c)} . E_j(s)$ follow. This proves (1) and (2).

If $s = 0$ then the constant term is, on $\mathfrak{S}_{i,t}$, a linear combination of h_{P_i} and $h_{P_i} \ln h_{P_i}$, which are bounded by $h_{P_i}^{1+\varepsilon}$ for any $\varepsilon > 0$, and (4) follows. Of course, if $d > 0$ (i.e., if B is not on the imaginary axis) then we may dispense with ε (i.e., with $d' > d$) in (4).

11.17 We conclude Section 11 with an example, the constant term in the most classical case, where $\Gamma = \mathrm{SL}_2(\mathbb{Z})$ and the Eisenstein series is right K-invariant. Then ∞ is the only cuspidal point and P is the standard parabolic subgroup P_0. We have

$$(1) \qquad \Gamma_P = CG.\Gamma_N, \qquad \Gamma_N = \left\{ \begin{pmatrix} 1 & m \\ 0 & 1 \end{pmatrix} \,\middle|\, m \in \mathbb{Z} \right\}.$$

A right coset $\Gamma_P \backslash \Gamma$ different from the identity is a set of all matrices

$$\begin{pmatrix} * & * \\ c & d \end{pmatrix} \in \mathrm{SL}_2(\mathbb{Z}),$$

where $c, d \in \mathbb{Z}$ are fixed, $(c, d) = 1$, and $c \geqslant 1$. The Eisenstein series to be considered is 10.3(3), and can be written

$$(2) \qquad E(s, z) = y^{(1+s)/2} + y^{(1-s)/2} \sum_{(c,d)=1, \, c \geqslant 1} |cz + d|^{-(1+s)}.$$

Here and in the sequel it is understood that c and d are integers. The constant term has the form

$$(3) \qquad E(s, z)_P = h_P^{1+s} + c(s) h_P^{1-s} = y^{(1+s)/2} + c(s) y^{(1-s)/2}.$$

We claim that

$$(4) \qquad c(s) = \pi^{1/2} \frac{\zeta(s)}{\zeta(s+1)} \frac{\Gamma(s/2)}{\Gamma((1+s)/2)},$$

where $\zeta(s)$ is Riemann's zeta function and $\Gamma(s)$ the Γ-function.

Proof. From (2) and (3) we obtain

$$(5) \qquad E(s, z)_P = y^{(1+s)/2} + y^{(1-s)/2} \cdot \sum_{\substack{(c,d)=1 \\ c \geqslant 1}} \int_0^1 \frac{dx}{|cz + d|^{s+1}}.$$

We have

$$\sum_{\substack{(c,d)=1 \\ c \geqslant 1}} \int_0^1 \frac{dx}{|cz + d|^{s+1}} = \sum_{\substack{(c,d)=1 \\ c \geqslant 1}} \frac{1}{c^{s+1}} \int_0^1 \frac{dx}{|z + d/c|^{s+1}}$$

$$= \sum_{\substack{(c,d)=1 \\ c \geqslant 1, \, d \bmod c}} \frac{1}{c^{s+1}} \sum_{q \in \mathbb{Z}} \int_0^1 \frac{dx}{|z + q + d/c|^{s+1}}$$

$$= \sum_{\substack{(c,d)=1 \\ c \geqslant 1, \, d \bmod c}} \frac{1}{c^{s+1}} \int_{-\infty}^{\infty} \frac{dx}{|z + d/c|^{s+1}},$$

whence

(6)
$$\sum_{\substack{(c,d)=1 \\ c \geqslant 1}} \int_0^1 \frac{dx}{|cz+d|^{s+1}} = \sum_{c \geqslant 1} \frac{\Phi(c)}{c^{s+1}} \int_{-\infty}^\infty \frac{dx}{|z|^{s+1}},$$

where Φ is Euler's function. We have

$$\int_{-\infty}^\infty \frac{dx}{|z|^{s+1}} = \int_{-\infty}^\infty \frac{dx}{(x^2+y^2)^{(s+1)/2}} = \frac{y}{y^{1+s}} \int_{-\infty}^\infty \frac{dx}{(1+x^2)^{(s+1)/2}},$$

so that

(7) $$E(s,z)_P = y^{(1+s)/2} + y^{(1-s)/2} \left(\sum_{c \geqslant 1} \frac{\Phi(c)}{c^{s+1}} \right) \int_{-\infty}^\infty \frac{dx}{(1+x^2)^{(s+1)/2}};$$

(4) now follows from (7) and the known relations

(8)
$$\sum_{c \geqslant 1} \frac{\Phi(c)}{c^{s+1}} = \frac{\zeta(s)}{\zeta(s+1)}$$

(see [27, p. 249, Thm. 288]) and

(9)
$$\int_{-\infty}^\infty \frac{dx}{(1+x^2)^{(s+1)/2}} = \pi^{1/2} \frac{\Gamma(s/2)}{\Gamma((1+s)/2)}$$

(see e.g. [26, p. 182, Formel 19]).

Our complex variable has been normalized so that the functional equation relates values at s and $-s$. In the classical literature, it is more usual to use the exponent of y in the constant term, say t. Then the critical line is at $\mathcal{R}t = 1/2$ and the functional equation relates the values at t and $1-t$. Set then $t = (1+s)/2$. Then

(10) $$E(t,z)_P = y^t + y^{1-t} \cdot \tilde{c}(t)$$

with $\tilde{c}(t) = c(2t-1)$, that is,

(11)
$$\tilde{c}(t) = \pi^{1/2} \frac{\zeta(2t-1) \cdot \Gamma(t-(1/2))}{\zeta(2t)\Gamma(t)}.$$

We know (11.11) that the analytic continuation of $E(s,z)$ is equivalent to that of $c(s)$ and that $E(s,z)$ and $c(s)$ have the same poles. The equality (4) shows that the analytic continuation of $E(s,\cdot)$ or of $c(s)$ is equivalent to that of the zeta-function (the one of $\Gamma(s)$ being taken for granted). The known properties of $\Gamma(s)$ and $\zeta(s)$ imply that, on $\mathcal{R}s \geqslant 0$, the function $E(s,z)$ is holomorphic except for a simple pole at $s = 1$. This will also follow from the results of Sections 12 and 13, which, incidentally, also yield the analytic continuation of $\zeta(s)$.

We have invoked 11.11 for the equivalence of the analytic continuation of $c(s)$ and of $E(s,z)$. However, in this special case, this equivalence can be seen from an explicit expansion of the series $E(s,z)$ in terms of the zeta function and of Bessel functions (see e.g. [36, 3.29] or [48, 1.4.12]).

12 Eisenstein series and automorphic forms orthogonal to cusp forms

Our first main goal here is to conclude the discussion of the analytic continuation of Eisenstein series by proving 11.13. To this effect we use formulas for scalar products that will also play an important role in Sections 16 and 17. The first formula is the scalar product of a fast decreasing E-series and an Eisenstein series (12.3, 12.6), from which we deduce that $\overline{C(s)} = {}^tC(\bar{s})$ as well as the unitarity – and hence boundedness and holomorphy – of $C(s)$ on the imaginary axis (12.7). The second is the scalar product of two truncated Eisenstein series (12.10) (the "Maass–Selberg relations"), from which the assertions on the poles in $\mathcal{R}s > 0$ follow easily (12.11, 12.12, 12.13). The remaining part of this section, from 12.15 on, is concerned with the space of automorphic forms of a given type that are orthogonal to cusp forms; the main results are 12.17 and 12.24. To establish them, we use two variants of the Maass–Selberg relations (12.16, 12.22) and some refinements of 11.16, proving that some estimates of moderate growth or fast decrease for certain functions parameterized by s varying in a compact subset of $\mathbb{C}$ are independent of s (12.19, 12.26).

12.0 As in 10.12, $\delta = 0, 1$. Let ε be the nontrivial element of CG. Given a closed subgroup H of G, we let

(1) $\qquad {}_\delta C(H\backslash G) = \{f \in C(H\backslash G), \ f(\varepsilon x) = (-1)^\delta f(x), \ x \in G\}.$

Then

(2) $\qquad\qquad C(H\backslash G) = {}_0C(H\backslash G) \oplus {}_1C(H\backslash G).$

The following simple remarks should be kept in mind.

(3) If $f \in {}_\delta C(H\backslash G)$ is of right K-type m and not zero, then $m \equiv \delta \bmod 2$.
(4) Let (P, A) be a cuspidal p-pair and $N = N_P$. Then ${}_\delta C(\Gamma_P.N\backslash G) \neq 0$ implies $\delta = 0$ or $\Gamma_P = \Gamma_N$.

12.1 Let P be a Γ-cuspidal parabolic subgroup and $f \in {}_\delta C(\Gamma_P.N\backslash G)$. The E-series E_f or $_PE_f$ is, by definition,

(1) $\qquad\qquad E_f(x) = \sum_{\gamma \in \Gamma_P\backslash\Gamma} f(\gamma.x).$

We shall consider such series only when they converge absolutely and locally uniformly. A main example is the Eisenstein series $E(P, s)$ for $\mathcal{R}s > 1$ (see 10.4). We shall have to use another type of E-series, which are compactly supported mod Γ.

Under these convergence conditions, we can permute summation and integration over K. Hence, in the notation of 2.20,

(2) $$(E_f)_m = E_{f_m} \quad (m \in \mathbb{Z}).$$

In particular, E_f is right K-finite if f is so.

12.2 Proposition. *Let $f \in {}_\delta C_c(\Gamma_P.N\backslash G)$. Then E_f converges absolutely and locally uniformly to an element of ${}_\delta C_c(\Gamma \backslash G)$, which is smooth if f is so.*

Proof. We show more strongly that there are only finitely many γ mod Γ_P such that $f(\gamma.x)$ is not identically zero on G.

There exists a compact set Q such that supp $f \subset \Gamma_P.Q$. Let (P', A') be a cuspidal p-pair and $\mathfrak{S}'$ a Siegel set with respect to (P', A'). The set of $\gamma \in \Gamma$ such that $\gamma(\mathfrak{S}') \cap Q \neq \emptyset$ is finite, by 3.16. There are thus only finitely many classes $\Gamma_P.\gamma$ such that $f(\gamma.x) \neq 0$ for some $x \in \mathfrak{S}'$, hence only finitely many terms in the series E_f that are not identically zero on $\mathfrak{S}'$. Because Γ has a fundamental set consisting of finitely many Siegel sets (and a compact set if we insist that the Siegel sets be disjoint (3.17)), our assertion is proved.

Remark. Assume f to be of K-type m on the right. Then (12.0), f not identically zero implies that either m is even or $\Gamma_P = \Gamma_N$. Using the Iwasawa decomposition, we see that f can be written as $f = u.\varphi_{P,m}$, where $u \in C_c(A)$ is viewed as a left N-invariant and right K-invariant function on G and $\varphi_{P,m}$ is as in 10.2. These are the f, with $u \in C_c^\infty(A)$, that will be mostly used here, as is the case in [31]. Here φ_P is normalized. We shall also write E_u or, if P and/or m need be specified, ${}_P E_u$ or ${}_P E_{u,m}$ for the function E_f in 12.2. Following Godement [24], we shall sometimes call E_f an *incomplete theta series*.

12.3 Proposition. *Let P, P' be Γ-cuspidal parabolic subgroups and E_f, $E_{f'}$ two E-series associated to P and P', respectively. Assume that E_f is fast decreasing and $E_{f'}$ has moderate growth. Then*

(1) $$(E_f, E_{f'})_{\Gamma\backslash G} = (E_{f,P'}, f')_{N'\backslash G} = (f, E_{f',P})_{N\backslash G}.$$

The proof of the equality of the first term and of either the second or third term is the same as that of 10.10 and is left to the reader.

12.4 Fourier–Laplace transform. We shall use the Fourier transform and, in particular, the Paley–Wiener theorem. Our notation will be somewhat different from the standard one, so here we recall and complete our conventions. Let (P, A)

be a (normal, as usual) p-pair. Recall (2.8) that we have identified A with the multiplicative group of strictly positive real numbers $\mathbb{R}^{*+}$ by means of the function $a \mapsto h_P(a)$, and $\mathbb{C}$ with the character group of A by assigning to $s \in \mathbb{C}$ the ($\mathbb{C}^*$-valued) character $a \mapsto h_P(a)^s$. Let $\mathfrak{a}$ be the Lie algebra of A, and let $\exp: \mathfrak{a} \to A$ be the exponential map. We identify $\mathfrak{a}$ with $\mathbb{R}$ so that the map $\mathbb{R} \to \mathbb{R}^{*+}$ defined by $\exp$ is the usual exponential. Then the interval $[-R, R]$ $(R > 0)$ on $\mathfrak{a}$ maps to the interval $[e^{-R}, e^R]$ on A. The Haar measure da becomes $dt.t^{-1}$ on $\mathbb{R}^{*+}$ and corresponds to the Lebesgue measure dx on $\mathbb{R}$. For $u \in C_c^\infty(A)$, let

$$(1) \qquad \hat{u}(s) = (2\pi)^{-1/2} \int_A u(a) h_P(a)^{-(1+s)} \, da$$

be its Fourier–Laplace transform. (The $1+$ is a normalization, so that the critical line will be $\mathcal{R}s = 0$. Note also that we omit the traditional i; i.e., we permute the roles of the real and imaginary parts of s.) As before, we write $s = \sigma + i\tau$ $(\sigma, \tau \in \mathbb{R})$. Conversely, for any $\sigma \in \mathbb{R}$ we have

$$(2) \qquad u(a) = (2\pi)^{-1/2} \int_{\sigma+i\mathbb{R}} \hat{u}(s) h_P(a)^{1+s} \, d\tau,$$

as follows from the standard Fourier inversion formula. Explicitly, let $\mathcal{F}$ and $\tilde{\mathcal{F}}$ be the usual Fourier transform and inverse Fourier transform, so that

$$\mathcal{F} \circ \tilde{\mathcal{F}} = \tilde{\mathcal{F}} \circ \mathcal{F} = \mathrm{Id}.$$

Then, by definition,

$$\hat{u}(s) = \mathcal{F}(u.h_P^{-1-\sigma})$$

and the right-hand side of (2), at a, is equal to $h_P(a)^{1+\sigma} \tilde{\mathcal{F}}(\hat{u})(a)$, and hence to

$$h_P(a)^{1+\sigma} \tilde{\mathcal{F}}\mathcal{F}(u.h_P^{-(1+\sigma)})(a) = h_P(a)^{1+\sigma}.u(a).h_P(a)^{-1-\sigma} = u(a).$$

The set of all $\hat{u}$ is characterized by the *Paley–Wiener theorem*: A function v on $\mathbb{C}$ is the Fourier–Laplace transform of $u \in C_c^\infty(A)$ with support in $[r^{-1}, r]$ $(r > 1)$ if and only if v is entire and, for each $n \in \mathbb{N}$, there exists a constant C_n such that

$$(3) \qquad |v(s)| \leqslant C_n.r^{|\sigma|}(1 + |s|)^{-n} \quad (s \in \mathbb{C}).$$

In particular, v is fast decreasing on any vertical strip $c \leqslant \sigma \leqslant c'$ $(c, c' \in \mathbb{R}, c \leqslant c')$ and its restriction to any vertical line $\sigma + i\mathbb{R}$ is square integrable (cf. e.g. [61, §VI.4] or [46, IX.3]). Such a function will be called a Paley–Wiener or PW function.

12.5 We now use the Fourier–Laplace transform to put the scalar products in 12.3(1) in a form that is basic for the sequel. We let (P', A') be a cuspidal p-pair. Let

$$(1) \qquad u \in C_c^\infty(A), \qquad v \in C(A').$$

Then E_u is absolutely and locally uniformly convergent to a compactly supported smooth function on $\Gamma \backslash G$ (12.2). We assume that E_v is absolutely and locally uniformly convergent and has moderate growth on $\Gamma \backslash G$. We claim first that

$$(2) \qquad E_u(x) = (2\pi)^{-1/2} \int_{\sigma+i\mathbb{R}} \hat{u}(s).E(P, s)(x)\, d\tau \quad (\sigma > 1).$$

Proof. By definition and (12.4(2)), we have

$$(3) \qquad E_u(x) = \sum_{\gamma \in \Gamma_P \backslash \Gamma} u(\gamma.x).\varphi_P(\gamma.x)$$

$$= \sum_{\gamma} \varphi_P(\gamma.x)(2\pi)^{-1/2} \int_{\sigma+i\mathbb{R}} \hat{u}(s).h_P(\gamma.x)^{1+s}\, d\tau.$$

The Eisenstein series

$$(4) \qquad E(P, s)(x) = \sum_{\gamma} \varphi_{P,s}(\gamma.x) = \sum_{\gamma} \varphi_P(\gamma.x)h_P(\gamma.x)^{1+s}$$

has a constant absolute majorant for all $s \in \sigma_0 + i.\mathbb{R}$ by 10.4. Hence it is uniformly convergent for $s \in \sigma + i.\mathbb{R}$, so that we can permute $\sum$ and $\int$ and this yields (2).

In the sequel, we consider two cases: either $P = P'$ or P and P' are not Γ-conjugate. Let $\delta_{P,P'}$ be equal to one in the first case and to zero in the second.

12.6 Proposition. *Let E_u and E_v be as above.*

(i) *Assume that $E(P', \cdot)$ is holomorphic at s. Then*

$$(1) \qquad (E_u, E(P', s))_{\Gamma \backslash G} = (2\pi)^{1/2}\{\delta_{P,P'}\hat{u}(-\bar{s}) + \overline{c_{P'|P}(s)}\hat{u}(\bar{s})\}.$$

(ii) *Assume that $v \in C_c^\infty(A')$. Then*

$$(2) \qquad (E_u, E_v)_{\Gamma \backslash G} = \int_{\sigma+i\mathbb{R}} d\tau\, \hat{u}(s)\{\delta_{P,P'}\overline{\hat{v}(-\bar{s})} + c_{P|P'}(s)\overline{\hat{v}(\bar{s})}\} \quad (\sigma > 1).$$

(iii) *Assume that $E(P', s)$ has a simple pole at $z \in (0, 1]$. Then*

$$(3) \qquad (E_u, \mathrm{Res}_{s=z} E(P', s))_{\Gamma \backslash G} = (2\pi)^{1/2}\hat{u}(z)\, \mathrm{Res}_{s=z} c_{P'|P}(s).$$

Proof. (i) We first assume that $\sigma = \mathcal{R}s > 1$. Then $E(P', s)$ satisfies the conditions imposed on E_v in 12.5. In the notation of 11.2, and since $\varphi_{P,s} = \varphi_{P,m}.h_P^{1+s}$, we have

$$(4) \qquad E(P', s)_P = \varphi_{P,m}(\delta_{P',P}h_P^{1+s} + c_{P'|P}(s)h_P^{1-s}).$$

By 12.3, $(E_u, E(P', s))_{\Gamma \backslash G}$ may be written as an integral over $N \backslash G$. We have $N \backslash G \cong A \times K$, and the measure on $A \times K$ induced from the Haar measure on G

is $h_P(a)^{-2}.da.dk$ (see 2.9). All the functions on the right-hand side are either right K-invariant or of right K-type m. In particular, $\varphi_{P,m}(nak) = \chi_m(k)$. In the integrand under consideration, each term involves $\varphi_{P,m}$ and $\bar{\varphi}_{P,m}$, so that the integral over K is one. If we take this into account, from 12.3 we have

$$(E_u, E(P', s))_{\Gamma\backslash G} = \int_A da\, h_P(a)^{-2} u(a)\{\delta_{P',P} h_P(a)^{1+\bar{s}} + c_{P'|P}(s).h_P(a)^{1-\bar{s}}\}$$

$$(5)\quad (E_u, E(P', s))_{\Gamma\backslash G} = \int_A da\, u(a)\{\delta_{P',P} h_P(a)^{-1+\bar{s}} + \overline{c_{P'|P}(s)} h_P(a)^{-1-\bar{s}}\}.$$

Then (i) follows in this case from the definition of $\hat{u}$ (12.4).

We can also write (1) as

$$(6)\qquad (E(P', s), E_u)_{\Gamma\backslash G} = (2\pi)^{1/2}\big(\delta_{P',P}\overline{\hat{u}(-\bar{s})} + c_{P'|P}(s).\overline{\hat{u}(\bar{s})}\big).$$

By 11.9, we can find a relatively compact open subset U of $\mathbb{C}$ that contains s, as well as a non-empty open subset of $\mathcal{R}s > 1$ on which $E(P', s)$ is holomorphic. The right-hand side of (6) is clearly holomorphic on U. Since E_u is compactly supported, so is the left-hand side of (6) by 11.10. Hence (i) in general follows by analytic continuation from the case $\mathcal{R}s > 1$.

(ii) From 12.5(2) we get

$$(E_u, E_v)_{\Gamma\backslash G} = (2\pi)^{-1/2}\left(\int_{\sigma+i\mathbb{R}} d\tau\, \hat{u}(s)\, E(P, s), E_v\right)_{\Gamma\backslash G}.$$

We claim that we can permute $\int$ and $(\cdot, E_v)_{\Gamma\backslash G}$. Indeed, E_v is compactly supported, $E(P, s)$ has an absolute termwise majorant on $\sigma + i\mathbb{R}$ that is of moderate growth (10.4), and $\hat{u}(s)$ is fast decreasing. Therefore,

$$(7)\qquad (E_u, E_v)_{\Gamma\backslash G} = (2\pi)^{-1/2}\int_{\sigma+i\mathbb{R}} d\tau\, \hat{u}(s)(E(P, s), E_v)_{\Gamma\backslash G}$$

and (ii) follows from (6), with u, v and P, P' interchanged.

(iii) By 11.10, the function $s \mapsto ((s - z)E(P', s), E_v)_{\Gamma\backslash G}$ is holomorphic in some disc U centered at z. By (6),

$$(8)\quad ((s - z)E(P', s), E_u)_{\Gamma\backslash G} = (2\pi)^{1/2}(s - z).\{\delta_{P'|P}\overline{\hat{u}(-\bar{s})} + c_{P'|P}(s)\overline{\hat{u}(\bar{s})}\}$$

on $U - \{z\}$. Then (iii) follows from (8) by letting $s \to z$.

12.7 Proof of 11.13(i). By the same computation as before, we have

$$(E_u, E_v)_{\Gamma\backslash G} = \overline{(E_v, E_u)}_{\Gamma\backslash G} = \left(\int_{\sigma+i\mathbb{R}} \hat{v}(s)(E(P', s), E_u)_{\Gamma\backslash G}\, d\tau\right)^{-} \qquad (\sigma > 1)$$

(where the superscript $^{-}$ on the right means complex conjugate) and, using (6),

$$\overline{(E_v, E_u)}_{\Gamma \backslash G} = \int_{\sigma+i\mathbb{R}} d\tau . \overline{\hat{v}(s)} \{\delta_{P,P'} \hat{u}(-\bar{s}) + \overline{c_{P'|P}(s)} \hat{u}(\bar{s})\},$$

which, after replacing s by $-\bar{s}$ for the first integral and by $\bar{s}$ for the second, yields

(1) $(E_v, E_u)_{\Gamma \backslash G} = \int_{\sigma+i\mathbb{R}} \hat{u}(s) \{\delta_{P,P'} \overline{\hat{v}(-\bar{s})} + \overline{c_{P',P}(\bar{s})} \overline{\hat{v}(\bar{s})}\} d\tau$ $(\sigma > 1).$

The relations (1) and 12.6(2) imply that

(2) $\displaystyle\int_{\sigma+i\mathbb{R}} d\tau \, c_{P|P'}(s) \hat{u}(s) \hat{v}(\bar{s}) = \int_{\sigma+i\mathbb{R}} d\tau \, \overline{c_{P'|P}(\bar{s})} \hat{u}(s) . \hat{v}(\bar{s}).$

Equality (2) is valid for all $u \in C_c^\infty(A)$ and $v \in C_c^\infty(A')$, whence

(3) $$c_{P|P'}(s) = \overline{c_{P'|P}(\bar{s})}$$

at first for $\mathcal{R}s > 1$. By analytic continuation, this is then valid at all points s such that $c_{P|P'}$ and $c_{P'|P}$ are holomorphic at s and $\bar{s}$.

We now return to the notation of Section 11, letting $P = P_i$ and $P' = P_j$ and hence $c_{P|P'} = c_{i,j}$ and $\delta_{P,P'} = \delta_{i,j}$. We then have

(4) $$c_{i,j}(\bar{s}) = \overline{c_{j,i}(s)},$$

at least at all points s such that the matrix $C(s) = (c_{i,j}(s))$ is regular at s and $\bar{s}$ and so outside a discrete subset D of $\mathbb{C}$. Outside D, (4) yields

(5) $$C(s) = {}^t \overline{C(\bar{s})}.$$

This is valid in particular on an open dense set of the imaginary axis. There, $\bar{s} = -s$ and the functional equation $C(s).C(-s) = \mathrm{Id}$, combined with (5), gives

(6) $$C(s).{}^t \overline{C(s)} = 1;$$

that is, $C(s)$ is unitary on the imaginary axis outside the discrete set $D \cap i\mathbb{R}$. In particular, $c_{i,j}(s)$ is bounded on $i\mathbb{R} - D$. Because it is meromorphic (by 11.9), $c_{i,j}(s)$ is then bounded and holomorphic on the whole imaginary axis, and (6) is valid on $i\mathbb{R}$, proving 11.13(i).

12.8   We next turn to the *Maass–Selberg relations*. In their general form, they give the scalar product of the truncations of two automorphic forms that are eigenfunctions of the Casimir operator. The most important case for our purposes is that of Eisenstein series, which we consider first.

We choose $T > 0$ large enough so that the Siegel sets $\mathfrak{S}_{i,T}$ ($1 \leqslant i \leqslant l$) at the cusps are disjoint and so that $\gamma.\mathfrak{S}_{i,T} \cap \mathfrak{S}_{i,T} \neq \emptyset$ implies $\gamma \in \Gamma_{N_i}$ (3.16). Recall that $\Gamma \backslash G$ is the union of the images of the $\mathfrak{S}_{i,T}$, which are mapped injectively modulo Γ_{N_i} onto their images in $\Gamma \backslash G$, and of a compact set (3.17). Let f be an automorphic form. Instead of using $\Lambda^T f$ to denote the effect of the truncation operator on f as in 11.6, we shall write f^T, as is often done. We let χ_j^T be the characteristic function of $\mathfrak{S}_{j,T}$ $(1 \leqslant j \leqslant l)$. By definition (11.6),

$$(1) \qquad\qquad f^T = f - \sum_j \chi_j^T . f_{P_j}.$$

Truncation has the following elementary properties:

$$(2) \qquad\qquad \chi_j^T .(f^T)_{P_j} = 0 \quad (1 \leqslant j \leqslant l)$$
$$(3) \qquad\qquad (f^T, g^T) = (f^T, g) = (f, g^T),$$

where g is also an automorphic form.

Proof. The relation (2) follows from the definition and the fact that $(f_P)_P = f_P$ (7.1(3)). The second equality in (3) was proved in 11.6; to establish the first, it suffices to show that

$$(4) \qquad\qquad (f - f^T, g^T) = 0.$$

We have

$$(f - f^T, g^T) = \sum_j (\chi_j^T . f_{P_j}, g^T).$$

Now $\chi_j^T f_{P_j}$ is zero outside $\mathfrak{S}_{j,T}$, so we can view $(\chi_j^T f_{P_j}, g^T)$ as an integral over $\Gamma_{N_j} \backslash N_j . A_j . K$ with respect to the measure $h_{P_j}(a)^{-2} \, dn \, da \, dk$. Performing first the integration over $\Gamma_{N_j} \backslash N_j$ yields

$$\int_{A_j \times K} \chi_j^T(ak) f_{P_j}(ak) \overline{(g^T)}_{P_j}(ak) h_P^{-2}(a) \, da \, dk,$$

which vanishes in view of (2).

12.9   Before going over to the Maass–Selberg relation, we insert two remarks to prepare for its proof.

(a) Let $p: G \to \Gamma \backslash G$ be the canonical projection. We have already pointed out that the induced projection $\Gamma_{N_i} \backslash N_i . A_i . K \to \Gamma \backslash G$ is injective on $\mathfrak{S}_{i,T}$. Therefore, if f is a Γ_{N_i}-invariant function on $\mathfrak{S}_{i,T}$ then

$$(1) \qquad \int_{\mathfrak{S}_{i,T}} \chi_i^T . f(g) \, dg = \int_{\Gamma_{N_i} \backslash N_i . A_i . K} \chi_i^T . f . h_{P_i}^{-2}(a) \, dn \, da \, dk$$

$$= \int_{A_i \times K} \chi_i^T . f_{P_i} h_{P_i}^{-2}(a) \, da \, dk.$$

If, moreover, f is right K-invariant, then

$$(2) \qquad \int_{\mathfrak{S}_{i,T}} \chi_i^T f \, dg = \int_{A_i} \chi_i^T(a) f_{P_i}(a) h_{P_i}^{-2}(a) \, da$$

and, if we take on A the coordinate $t = h_P(a)$, then χ_i^T is the characteristic function of $[t, \infty)$ and

(3)
$$\int_{\mathfrak{S}_{i,T}} \chi_i^T f_{P_i} \, dg = \int_T^\infty f_{P_i}(t) t^{-3} \, dt.$$

(b) We have assumed $\gamma(\mathfrak{S}_{i,T}) \cap \mathfrak{S}_{i,T} = \emptyset$ if $\gamma \notin \Gamma_{N_i}$. Therefore, if f is Γ-invariant on the left then $(\chi_i^T . f)(\gamma x) = 0$ for $\gamma \notin \Gamma_{N_i}$ and $x \in \mathfrak{S}_{i,T}$, and we can also write the truncation operator as

(4)
$$f^T(x) = f(x) - \sum_i \sum_{\gamma \in \Gamma_{P_i} \backslash \Gamma} (\chi_i^T . f_{P_i})(\gamma x) \quad (x \in G).$$

This is the definition to be adopted in more general cases, where the preceding separation condition is not met. The sum is always finite by the Siegel property.

In the sequel, $m \in \mathbb{Z}$ is fixed and we write E_i or $E_i(s)$ for $E(P_i, m, s)$. It is of course also a function on G, but the dependence on $x \in G$ need not be made explicit until we integrate over the A_k.

12.10 Theorem. *Let $r, s \in \mathbb{C}$ be such that E_i (resp. E_j) is holomorphic at r (resp. s). Assume that $r \neq \pm \bar{s}$ and $r, s \neq 0$. Then*

(1)
$$(E_i^T(r), E_j^T(s)) = \delta_{i,j} \frac{T^{r+\bar{s}}}{r+\bar{s}} + \frac{T^{\bar{s}-r}}{\bar{s}-r} c_{i,j}(r)$$
$$+ \frac{T^{r-\bar{s}}}{r-\bar{s}} \overline{c_{j,i}(s)} - \frac{T^{-r-\bar{s}}}{r+\bar{s}} \sum_k c_{ik}(r) \overline{c_{jk}(s)}.$$

Proof. By analytic continuation, it suffices to prove (1) under the conditions

(2)
$$\mathcal{R}r > \mathcal{R}s > 1,$$

which we assume. By 11.6 and 12.8(3), the left-hand side is equal to $(E_i^T(r), E_j(s))$, hence

(3)
$$(E_i^T(r), E_j^T(s)) = \left(E_i(r) - \sum_k \chi_k^T . E_i(r)_{P_k}, E_j(s) \right).$$

This also shows that the left-hand side of (1) is finite, since $E_i^T(r)$ is fast decreasing on Siegel sets and $E_j(s)$ has moderate growth.

For $k \neq i$, let

(4)
$$M_k = (\chi_k^T . E_i(r)_{P_k}, E_j(s)).$$

This can also be written as

$$M_k = (\chi_k^T . E_i(r)_{P_k}, E_j(s))_{\Gamma_P \backslash G}.$$

Using the remark within 10.10, 11.3(ii), and 12.9(a), we have

$$M_k = (\chi_k^T . E_i(r)_{P_k}, E_j(s)_{P_i})_{N_k \backslash G}$$

$$= \int_{A_k} h_{P_k}^{-2}(a) \chi_k^T(a) c_{i,k}(r) h_{P_k}^{1-r}(a) \{ \delta_{j,k} h_{P_k}^{1+\bar{s}}(a) + \overline{c_{j,k}(s)} h_{P_k}^{1-\bar{s}}(a) \} \, da.$$

Replace h_{P_k} by t and hence da by $dt.t^{-1}$. Then $\chi_k^T | A_k$ is the characteristic function of $[T, \infty)$, and we obtain

$$M_k = \int_T^\infty dt \, \{ \delta_{j,k} c_{i,k}(r) t^{\bar{s}-r-1} + c_{i,k}(r) \overline{c_{j,k}(s)} t^{-r-\bar{s}-1} \}.$$

Under our assumptions, $\mathcal{R}(\bar{s} - r)$ and $\mathcal{R}(-r - \bar{s})$ are strictly less than zero; therefore,

$$(5) \qquad M_k = -\delta_{j,k} \frac{T^{\bar{s}-r}}{\bar{s} - r} c_{i,k}(r) + \frac{T^{-r-\bar{s}}}{r + \bar{s}} c_{i,k}(r) \overline{c_{j,k}(s)}.$$

We recall that $k \neq i$ in (5). There remains to consider

$$(6) \qquad M_{i,j} = \left(E_i(r) - \chi_i^T . E_i(r)_{P_i}, E_j(s) \right).$$

By the definition of E_i and by 12.9(b), the value of the first argument at $x \in G$ can be written as

$$\sum_{\gamma \in \Gamma_{P_i} \backslash \Gamma} (\varphi_{i,m,r} - \chi_i^T E_i(r)_{P_i})(\gamma x).$$

As in 12.3, we have

$$(7) \qquad M_{i,j} = (\varphi_{i,m,r} - \chi_i^T . E_i(r)_{P_i}, E_j(s)_{P_i})_{N_i \backslash G}.$$

By the manipulations of 12.9(a) and since $\varphi_{i,m,r} = h_{P_i}^{1+r} \varphi_{P_i,m}$, we obtain

$$(8)$$
$$M_{i,j} = \int_{A_i} h_{P_i}^{-2}(a) \{ (1 - \chi_i^T(a)) h_{P_i}^{1+r}(a) - \chi_i^T(a) . c_{i,i}(r) h_{P_i}^{1-r}(a) \} \overline{E_j(s)}_{P_i}(a) \, da.$$

We have

$$(9) \qquad E_j(s)_{P_i} = \delta_{j,i} h_{P_i}^{1+s}(a) + c_{j,i}(s) h_{P_i}^{1-s}(a) \quad (a \in A_i).$$

We again switch to $t = h_{P_i}$. Then $1 - \chi_i^T$ is the characteristic function of $(0, T)$, hence

$$(10) \qquad M_{i,j} = \int_0^T dt \, (\delta_{j,i} t^{r+\bar{s}-1} + \overline{c_{j,i}(s)} t^{r-\bar{s}-1})$$

$$- \int_T^\infty dt \, c_{i,i}(r) (\delta_{j,i} t^{\bar{s}-r-1} + \overline{c_{j,i}(s)} t^{-r-\bar{s}-1})$$

$$(11) \qquad M_{i,j} = \frac{T^{r+\bar{s}}}{r + \bar{s}} \delta_{j,i} + \frac{T^{r-\bar{s}}}{r - \bar{s}} \overline{c_{j,i}(s)}$$

$$+ \frac{T^{\bar{s}-r}}{\bar{s} - r} \delta_{j,i} c_{i,i}(r) - \frac{T^{-r-\bar{s}}}{r + \bar{s}} c_{i,i}(r) \overline{c_{j,i}(s)}.$$

(Note that, by (2), both $\mathcal{R}(r + \bar{s})$ and $\mathcal{R}(r - \bar{s})$ are strictly greater than zero.)
By (3),

$$(E_i^T(r), E_j^T(s)) = M_{i,j} - \sum_{k \neq i} M_k.$$

Then (1) follows from (5) and (11).

Remark. As far as I know, the term "Maass–Selberg relations" was coined by
Harish-Chandra [31]. As pointed out by Selberg in [50], such formulas can be obtained from two slightly different points of view. The first derivation, by H. Maass
[42] uses the Green formula, as does the proof in the body of the text of [50, p. 651,
7.41]. In Remark 3 [50, pp. 672–73], Selberg points out that this was not his original proof, and sketches the latter. The one given here is essentially the same.

12.11 Proof of 11.13(ii), (iii).   (ii) We must show that the $E_j(s)$ and $c_{i,j}(s)$
have no pole on $\mathcal{R}s \geqslant 0$, except possibly for finitely many simple poles in $(0, 1]$.
It suffices to consider the $c_{i,j}(s)$ (11.11), which we already know are holomorphic
on the imaginary axis (12.7). So let $s_0 = \sigma + i\tau$, with $\sigma > 0$.

Let first $\tau \neq 0$. Fix j and assume to the contrary that some $c_{i,j}$ has a pole
at s_0, and let $m \geqslant 1$ be the maximum of the orders of the poles of the $c_{i,j}(s)$
$(i = 1, \ldots, l)$ at s_0. Let $r = s = s_0 + i.t$ with t real, not zero, and small enough
so that the $c_{i,j}$ are holomorphic at s and $\tau \pm t \neq 0$. We have

$$(1) \qquad s - \bar{s} = 2i(\tau + t) \neq 0, \qquad s + \bar{s} = 2\sigma > 0.$$

We can apply the Maass–Selberg relations for $i = j$ and $r = s$ to yield

$$(2) \qquad \frac{T^{2\sigma}}{2\sigma} + \mathcal{J}\left(\frac{T^{2i(t+\tau)}}{2i(t + \tau)} \overline{c_{j,j}(s)} \right) \geqslant \frac{T^{-2\sigma}}{2\sigma} \sum_k |c_{j,k}(s)|^2$$

(since the left-hand side of 12.10(1) is greater than or equal to zero).

Multiply both sides by t^{2m} and let $t \to 0$. Then the left-hand side of (2) tends
to zero, but the right-hand side has a strictly positive limit. This is a contradiction, which proves that the $c_{i,j}(s)$ and $E_j(s)$ have no pole outside the real axis on
$\mathcal{R}s > 0$.

Let now $\tau = 0$. Assume that at least one of the $c_{i,j}$ has a pole at $s_0 = \sigma$. Then
$\sigma \leqslant 1$. Let m be the maximum of the orders of those poles. The left-hand side
of (2) has at most a pole of order $m + 1$, whereas the right-hand side grows like
t^{-2m}. Assume, contrary to 11.13(ii), that $m > 1$. Multiply again both sides of (2)
by t^{2m} and let $t \to 0$. Then, as before, the left-hand side tends to zero while the
right-hand side has a strictly positive limit, whence a contradiction.

Assume now that $E_i(s)$ has a simple pole at $\sigma \in (0, 1]$, and let E'_i be its residue there. The latter is an automorphic form (11.9). To show that it is square integrable, it suffices to prove for each j that its constant term E'_{i,P_j} is square integrable on a Siegel set $\mathfrak{S}_j$ with respect to P_j (7.7). If $c_{i,j}$ is holomorphic at s_0 then this constant term is identically zero, so assume $c_{i,j}$ has a simple pole at σ. Then the constant term of E'_i at P_j is a multiple of $h_{P_j}^{1-\sigma}$. There remains to see that the latter is square integrable on $\mathfrak{S}_j$, which is immediate: set $t = h_P$. Then, up to a constant, the square norm of $t^{1-\sigma}$ on $\mathfrak{S}_j$ is $\int_c^\infty t^{-3} . t^{2-2\sigma} \, dt$ for some $c > 0$, which is finite since $\sigma > 0$.

This concludes the proof of 11.13(ii). To establish (iii) we may assume $R > 1$. By 11.13(i) and (ii), the function $c_{i,j}(s)$ is holomorphic in the region under consideration, including the boundary. By the maximum principle, it suffices to show that $|c_{i,j}(s)|$ is bounded on the boundary. On $\mathcal{I}s = R$, this follows from 11.3; on the segments $\mathcal{R}s \in [0, R]$ with $\mathcal{I}s = \pm T$ it follows from continuity, and on $\mathcal{R}s = 0$ from 11.13(i).

12.12 Corollary. *Let $\sigma \in (0, 1]$ and $j, k \in \{1, \ldots, l\}$.*

(i) *If $c_{j,k}(s)$ has a pole at σ, then so do $c_{j,j}(s)$ and $c_{k,k}(s)$. In particular, σ is a pole of $E_j(s)$ if and only if it is one for $c_{j,j}(s)$.*

(ii) *If E_j and E_k have a pole at σ, then $c'_{j,k}(\sigma) = (E'_j(\sigma), E'_k(\sigma))$, where $'$ denotes residues.*

Proof. (i) We specialize the Maass–Selberg relations to the case $r = s = \sigma + i\tau$ $(\sigma > 0, \tau \neq 0)$ and get

$$(1) \qquad (E_j(s)^T, E_k(s)^T) = \delta_{j,k} \frac{T^{2\sigma}}{2\sigma} - \frac{T^{-2\sigma}}{2\sigma} \sum_n c_{j,n}(s) \overline{c_{k,n}(s)}$$
$$+ \frac{T^{2i\tau}}{2i\tau} \overline{c_{k,j}(s)} - \frac{T^{-2i\tau}}{2i\tau} c_{j,k}(s).$$

Let first $j = k$; then (1) gives

$$(2) \qquad \|E_j(s)^T\|^2 + \frac{T^{-2\sigma}}{2\sigma} \sum_n |c_{j,n}(s)|^2 = \frac{T^{2\sigma}}{2\sigma} + \mathcal{I}\left(\frac{T^{2i\tau}}{2i\tau} \overline{c_{j,j}(s)}\right).$$

Let now $\tau \to 0$. If $c_{j,j}$ is holomorphic at σ, then the right-hand side grows at most like τ^{-1}. On the other hand, if one of the $c_{j,k}$ has a pole at σ, then the left-hand side grows at least like τ^{-2} – a contradiction. Together with the relation $c_{j,k}(\bar{s}) = \overline{c_{k,j}(s)}$, this proves the first assertion of (i) and, combined with 11.11, also the second one.

(ii) Assume that E_j and E_k have a pole at σ. Multiply both sides of (1) by τ^2 and let $\tau \to 0$. Noting that truncation and taking residues commute, we have

(3) $(E_j'(\sigma)^T, E_k'(\sigma)^T) = -\dfrac{T^{-2\sigma}}{2\sigma} \sum c_{j,n}'(\sigma) \overline{c_{k,n}'(\sigma)} + c_{j,k}'(\sigma),$

and (ii) follows by letting $T \to \infty$.

12.13 Proposition. *The Eisenstein series $E_j(0, s)$ has a pole at 1. Its residue is the constant function, equal to $\mathrm{vol}(\Gamma \backslash G)^{-1}$ ($j = 1, \dots, l$).*

Proof. Assume that $E_j(0, s)$ has a pole at 1. The constant term of $E_j'(0, 1)$ at the kth cusp is zero if $c_{j,k}(s)$ is holomorphic at 1, and equal to $c_{j,k}'(1).h_{P_k}^0 = c_{j,k}'(1)$ if not. Hence all the constant terms of $E_j'(0, 1)$ are constant. The formation of the constant term commutes with right translations; therefore, $r_g E_j'(0, 1)$ has the same constant terms as $E_j'(0, 1)$ ($g \in G$) and the difference $r_g E_j'(0, 1) - E_j'(0, 1)$ is a cusp form. But Eisenstein series or their residues are orthogonal to cusp forms (11.11), hence $E_j'(0, 1)$ is right-invariant under G and so is a constant, say c. Then $c = c_{j,k}'$ for all k. Since $c_{j,j}' \neq 0$ by assumption, $c_{j,k}' \neq 0$ for all k and 11.11 shows that $E_k(0, s)$ also has a pole at 1 ($k = 1, \dots, l$). By 12.12,

$$c = \|E_j'(0, 1)\|^2 = c^2 \, \mathrm{vol}(\Gamma \backslash G),$$

whence $c \, \mathrm{vol}(\Gamma \backslash G) = 1$.

To conclude the proof, there remains only to show that at least one of the $E_j(0, s)$ has a pole at 1. The constant function 1 is a square integrable automorphic form orthogonal to cusp forms and hence belongs to V_0. It is annihilated by C and so belongs to the discrete spectrum of C. By 13.8 (the proof of which does not use the present proposition), the latter is spanned by the residues of Eisenstein series and cusp forms. Hence there must exist an Eisenstein series with a pole at 1. To yield a constant function, it must be right K-invariant and hence be one of the $E_j(0, s)$.

12.14   In all the foregoing, the right K-type was understood. In principle, $C(s)$ depends on it. Let us write it here as $_mC(s)$, and also $_mc_{i,j}(s)$, $_mc_{P|P'}(s)$, to remind ourselves that we are dealing with constant terms of Eisenstein series $E(P, m, s)$. Then we have

(1) $_mC(s) = {}_{-m}^t C(s)$ $(m \in \mathbb{Z})$

for any regular value of $_mC$. In particular, $_0C(s)$, called the "scattering matrix" in [36, p. 94], is symmetric, as shown there.

The relation (1) is seen by computations very similar to those of 12.6 and 12.7 but applied to a symmetric bilinear form on functions on $\Gamma \backslash G$:

$$\langle u, v \rangle = \int_{\Gamma \backslash G} u(g).v(g) \, dg$$

rather than the sesquilinear form $(,)_{\Gamma \backslash G}$, assuming one factor to be of right K-type m and the other to be of right K-type $-m$. Again, in the proof of the analog of (1) in 12.6, namely

$$(2) \qquad \langle E_u, E(P', -m, s)\rangle = (2\pi)^{1/2}\{\delta_{P', P}\hat{u}(-s) + {}_{-m}c_{P'|P}(s)\hat{u}(s)\},$$

the integral over K is the one of $\chi_m(k).\chi_{-m}(k)$ and hence is equal to 1; the computation there carries over but without any complex conjugate. This gives (1). Then, instead of (2) in 12.6 we have

$$(3) \qquad \langle E_u, E_v\rangle = \int_{\sigma+i\mathbb{R}} d\tau\, \hat{u}(s)\{\delta_{P', P}\hat{v}(-s) + {}_m c_{P|P'}(s)\hat{v}(s)\}v,$$

where u (resp. v) is of right K-type m (resp. $-m$); or also, as in 12.7(1),

$$(4) \qquad \langle E_u, E_v\rangle = \int_{\sigma+i\mathbb{R}} d\tau\, \hat{u}(s)\{\delta_{P, P'}\hat{v}(-s) + {}_{-m}c_{P'|P}(s)\hat{v}(s)\},$$

and the comparison of (3) and (4) now yields (1) instead of 12.7(3).

12.15 Given $s \in \mathbb{C}$ and $m \in \mathbb{Z}$, we let $\mathcal{A}_1(s, m)$ be the space of automorphic forms of right K-type m, which are eigenfunctions of C with eigenvalue $(s^2 - 1)/2$ and are orthogonal to cusp forms. Let $f \in \mathcal{A}_1(s, m)$; then its constant term at P_i can be written as

$$(1) \qquad\qquad f_{P_i} = a_i.\varphi_{i,m,s} + b_i.\varphi_{i,m,s} \quad (s \neq 0),$$
$$(2) \qquad\qquad f_{P_i} = \varphi_{P_i, m}.h_{P_i}(a_i + b_i \ln h_{P_i}) \quad (s = 0)$$

(see 11.1(4) and 11.9(7), (8)). Given m, we let δ be the element of $\{0, 1\}$ that is congruent to $m \bmod 2$. If $\delta = 1$, then $f_P = 0$ for $i > l(\delta)$ (see 10.12(2)). Let

$$(3) \qquad\qquad a_f = (a_1, \ldots, a_{l(\delta)}), \qquad b_f = (b_1, \ldots, b_{l(\delta)}).$$

The map $f \mapsto (a_f, b_f)$ of $\mathcal{A}_1(s, m)$ into $\mathbb{C}^{2l(\delta)}$ is *injective*, because if f is in its kernel then it is a cusp form and hence orthogonal to itself. The space $\mathcal{A}_1(s, m)$ contains the Eisenstein series $E(s, m)$, which are holomorphic at s, and more generally the limits

$$\lim_{r \to s}(r - s)^{c(E)} E(r, m) \quad (c(E): \text{order of the pole of } E(r, m) \text{ at } s).$$

We want to prove that $\mathcal{A}_1(s, m)$ is generated by those functions if $s \neq 0$. The main point is to show that it is of dimension $\leqslant l(\delta)$, and this will follow from the Maass–Selberg relations extended to a slightly more general case than in 12.10.

Assume that all Eisenstein series $E_i(r, m)$ are holomorphic at s ($i \leqslant l(\delta)$). (By 11.11 and the adjointness relation 12.7(4), this is equivalent to all the $E_i(r, m)$ being holomorphic at $\bar{s}$.) In that case, for r close to s, a general Eisenstein series $E(r, m)$ is a linear combination

(4) $E(s, m) = \sum_j \mu_j . E_j(s, m) \quad (\mu_j \in \mathbb{C}, \; j = 1, \ldots, l(\delta)).$

Let

(5) $E(r, m)_{P_i} = u_i(r).\varphi_{i,m,r} + v_i(r).\varphi_{i,m,-r} \quad (r \neq 0).$

By 11.3, we have

(6) $u_i(r) = \mu_i, \quad v_i(r) = \sum_j \mu_j . c_{j,i}(r) \quad (i = 1, \ldots, l(\delta)).$

By 11.14, the $E_i(s, m)$ are linearly independent, hence $E(s, m)$ is completely characterized by the first of the two $l(\delta)$-vectors

(7) $u_{E(r)} = (u_1(r), \ldots, u_{l(\delta)}(r)), \qquad v_{E(r)} = (v_1(r), \ldots, v_{l(\delta)}(r)).$

We let (,) denote the standard Hermitian scalar product on $\mathbb{C}^l$.

12.16 Theorem. *We retain the previous notation. Let $r, s \neq 0$ and $r \pm \bar{s} \neq 0$. Assume that the $E_i(s, m)$ are holomorphic at r and let $f \in A_1(s, m)$. Then*

(1) $(r^2 - \bar{s}^2)(E(r, m)^T, f^T)$

$$= (r - \bar{s})T^{r+\bar{s}}(u_{E(r)}, a_f) + (r + \bar{s})T^{r-\bar{s}}(u_{E(r)}, b_f)$$

$$- (r + \bar{s})T^{\bar{s}-r}(v_{E(r)}, a_f) - (r - \bar{s})T^{-r-\bar{s}}(v_{E(r)}, b_f).$$

Proof. The proof is quite similar to that of 12.10, which is a special case. We indicate briefly how to modify it. It suffices to prove the theorem for

$$E(r, m) = E_i(r, m),$$

since the general case will then follow by linearity. Hence we need only go through the steps of the proof of 12.10, replacing $E_j(s, m)$ by f and hence $E_j(s, m)_{P_k}$ by f_{P_k}. As before,

(2) $(E_i(r, m)^T, f^T) = M_i - \sum_{k \neq i} M_k,$

where

(3) $M_k = (\chi_k^T . E_i(r)_{P_k}, f_{P_k})_{N_k \backslash G} \quad (k \neq i),$

(4) $M_i = (E_i(r, m) - \chi_i^T E_i(r, m)_{P_i}, f_{P_i})_{N_i \backslash G}.$

As in 12.10, we have

(5)
$$M_k = \int_T^\infty dt \, (c_{i,k}(r)\bar{a}_k t^{\bar{s}-r-1} + c_{i,k}(r)\bar{b}_k t^{-r-\bar{s}-1}),$$

(6)
$$M_k = -\frac{T^{\bar{s}-r}}{\bar{s}-r} c_{i,k}(r)\bar{a}_k + \frac{T^{-r-\bar{s}}}{r+\bar{s}} c_{i,k}(r)\bar{b}_k,$$

(7)
$$M_i = \int_0^T dt \, (\bar{a}_i t^{r+\bar{s}-1} + \bar{b}_i t^{r-\bar{s}-1})$$
$$- \int_T^\infty dt \, c_{i,i}(r)(\bar{a}_i t^{\bar{s}-r-1} + \bar{b}_i t^{r-\bar{s}-1}),$$

(8)
$$M_i = \frac{T^{r+\bar{s}}}{r+\bar{s}}\bar{a}_i + \frac{T^{r-\bar{s}}}{r-\bar{s}}b_i + \frac{T^{\bar{s}-r}}{\bar{s}-r}\bar{a}_i c_{i,i}(r) - \frac{T^{-r-\bar{s}}}{r+\bar{s}} c_{i,i}(r)\bar{b}_i.$$

We then multiply both sides of (6) and (8) by $\mu_i(r^2 - s^2)$, use 12.15(6), and sum over i. This yields (1).

12.17 Theorem. *Let $m \in \mathbb{Z}$ and $r, s \in \mathbb{C}^*$. Let $\mathcal{E}(r, m)$ be the subspace of $\mathcal{A}_1(s, m)$ generated by the Eisenstein series $E(r, m)$, which are holomorphic at r, and let $\mathcal{PE}(r, m)$ be the subspace of $\mathcal{A}_1(s, m)$ generated by the limits*

$$\lim_{r \to s}(r - s)^{c(E)} E(r, m) \quad (c(E): \text{order of the pole of } E(r, m) \text{ at } s).$$

Then $\mathcal{A}_1(s, m) = \mathcal{PE}(s, m)$ and has dimension $l(\delta)$ (see 10.12 for $l(\delta)$).

Proof. We first show that

(1)
$$\dim \mathcal{PE}(s, m) = l(\delta).$$

Fix a compact neighborhood U of s in $\mathbb{C}^*$ such that the $E_i(r)$ are holomorphic in U except possibly at s. The spaces $\mathcal{E}(r, m)$ and $\mathcal{E}(r', m)$ $(r, r' \in U - \{s\})$ are then canonically isomorphic and $l(\delta)$-dimensional (11.14). We assume that $z \in U$ and $z \neq s$, and want to establish an isomorphism of $\mathcal{E}(z, m)$ onto $\mathcal{PE}(s, m)$.

We let d be the maximum of the order of the poles of Eisenstein series at s. For $1 \leqslant j \leqslant d$, let $A_j \subset \mathcal{E}(r, m)$ be the space spanned by the values at z of Eisenstein series $E(r, m)$ $(r \in U)$ that have at s a pole of order $\leqslant j$. On A_j we consider the operator

$$\pi_j: E(r, m) \mapsto \lim_{r \to s}(r - s)^j . E(r, m),$$

and denote by C_j its image in $\mathcal{A}_1(s, m)$; its kernel is A_{j-1}. Let B_j be a supplement to A_{j-1} in A_j for $j \geqslant 1$, and set $B_0 = A_0$. Define $\pi_0: B_0 \to \mathcal{A}_1(s, m)$ as the map that associates $E(s, m)$ to $E(r, m)$, and let C_0 be its image. By 11.14(i), π_0 is an *isomorphism* of B_0 onto C_0. The map π_j $(j \geqslant 1)$ is, by construction, an isomorphism of B_j onto C_j. Therefore, if $E(r, m)$ is such that $E(z, m) \in B_j$ and is not zero, then $E(r, m)$ has a pole of order j (exactly) at s. From this, it follows easily that the spaces C_j $(0 \leqslant j \leqslant d)$ are linearly independent, whence

$$\dim \mathcal{PE}(s, m) = \sum_{j \geqslant 0} \dim C_j = \sum_{j \geqslant 0} \dim B_j = l(\delta).$$

This proves that $\dim \mathcal{A}_1(s, m) \geqslant l(\delta)$. The reverse inequality will follow from the Maass–Selberg relations. We use the notation of 12.15 and 12.16. Recall that $E(r, m)$ has a pole of order j at s if and only if j is the maximum of the orders of the poles of its constant terms at s (11.11). We use the same notation, except that now we consider poles at $\bar{s}$ rather than s, that is, U is now a neighborhood of $\bar{s}$. We may and do assume moreover that its closure is compact, and that the E_j are holomorphic on that closure, except for possible poles at $\bar{s}$. If $E(z, m) \in B_j$, let

$$(2) \qquad u_{E(\bar{s})}^{(j)} = \lim_{r \to \bar{s}} (r - \bar{s})^j . u_{E(r)}, \qquad v_{E(\bar{s})}^{(j)} = \lim_{r \to \bar{s}} (r - \bar{s})^j . v_{E(r)}.$$

Of course, $u_{E(\bar{s})}^{(j)} = 0$ if $j \geqslant 1$.

We use 12.16(1) for $E(r, m)$ and f. We claim first that

$$(r - \bar{s})^j . (E(r, m)^T, f^T)$$

is bounded on U. Indeed, by 12.8(3), the scalar product can also be written $(E(r, m), f^T)$. It follows from 11.16 that $(r - \bar{s})^j . E(r, m)(x)$ is bounded for $r \in U$ and x in a compact set, and is bounded by $q h_i^c(x)$ on a Siegel set $\mathfrak{S}_{i, T}$ with q, c independent of $r \in U$. Because f^T is fast decreasing on $\mathfrak{S}_{i, T}$, our assertion follows. Now multiply both sides of 12.16(1) by $(r - \bar{s})^j$ and let $r \to \bar{s}$. Then the left-hand side tends to zero and the right-hand side to

$$2\bar{s}(u_{E(\bar{s})}^{(j)}, b_f) - 2\bar{s}(v_{E(\bar{s})}^{(j)}, a_f).$$

Since $\bar{s} \neq 0$ by assumption, it follows that

$$(3) \qquad (u_{E(\bar{s})}^{(j)}, b_f) = (v_{E(\bar{s})}^{(j)}, a_f).$$

An element of $\mathcal{A}_1(s, m)$ is characterized by its constant terms, so the map

$$Q(\bar{s}, m) \mapsto (v_{F(\bar{s})}^{(j)}, u_{F(\bar{s})}^{(j)}) \qquad (Q(\bar{s}, m) \in C_j)$$

is a monomorphism of $\mathcal{PE}(\bar{s}, m)$ into $\mathbb{C}^{2l(\delta)}$. As a result, then, the vectors (a_f, b_f) $(f \in \mathcal{A}_1(s, m))$ are orthogonal to an $l(\delta)$-dimensional subspace of $\mathbb{C}^{2l(\delta)}$ and so form a space of dimension $\leqslant l(\delta)$. Since they characterize the elements of $\mathcal{A}_1(s, m)$, this provides the reverse inequality.

12.18   There remains the origin, where all Eisenstein series are holomorphic (12.7) and hence $\mathcal{PE}(0, m) = \mathcal{E}(0, m) \subset \mathcal{A}_1(0, m)$. We want to prove that $\mathcal{A}_1(0, m)$ has dimension $l(\delta)$, too. However, the functions $E_i(0, m)$ are not necessarily linearly independent, so a new argument is needed to show that $\mathcal{A}_1(s, m)$ has at least dimension $l(\delta)$. The reverse inequality will again follow from the Maass–Selberg relations, or rather from a slight variant valid for $s = 0$.

We first draw a consequence of the growth estimate 11.16 around the origin. In this case B is a compact neighborhood of the origin in $\mathbb{C}$ – at all points of which the $E_j(s)$ are holomorphic – that admits a compatible $\alpha \in I_c^\infty(G)$ (10.7). Also $M(s) = 1$. In 11.16, we may assume q_N and q so chosen that it is valid for all j.

On $\mathbb{C}^{l(\delta)}$, we let $|\cdot|$ be the norm defined by $|u| = \max_i |u_i|$ ($u = (u_1, \ldots, u_{l(\delta)})$). Let

$$E(s, m) = E(s) = \sum_j \mu_j . E_j(s) \quad (s \in \mathbb{C}, \ s = 1, \ldots, l(\delta))$$

be an arbitrary Eisenstein function. Because the $c_{j,i}$ are bounded on B, we see from 12.15(6) that there exists a constant c, independent of the μ_j, such that

$$|u_{E(s)}| = \max|\mu_i|, \quad |v_{E(s)}| \leqslant c.|u_{E(s)}| \quad (s \in B).$$

12.19 Corollary (of 11.16). *We keep the previous notation and let C, $\mathfrak{S}_{i,t}$, and d be as in 11.16. There exists a constant c such that $E(r, m)$ satisfies the following conditions.*

(1)
$$|E(s, m)(x)| \leqslant c|u_{E(s)}| \quad (s \in B, \ x \in C),$$

(2)
$$|E(s, m)(x)| \leqslant |u_{E(s)}|.h_{P_i}^{1+d}(x) \quad \left(x \in \mathfrak{S}_{i,t} \ (i = 1, \ldots, l(\delta)), \ s \in B - \{0\}\right).$$

(3) *Given $N \in \mathbb{N}$, there exists a constant δ_N such that*

$$|E(s, m)'| \leqslant \delta_N.|u_{E(s)}|h_{P_i}(x)^{-N} \quad \left(x \in \mathfrak{S}_{i,t} \ (i = 1, \ldots, l(\delta)), \ s \in B\right).$$

(4) *For any $d' > d$,*

$$|E_j(s)(x)| \leqslant ch_{P_i}(x)^{1+d'}|u_{E(s)}| \quad \left(x \in \mathfrak{S}_{i,t} \ (i = 1, \ldots, l(\delta)), \ s \in B\right).$$

Note that, by 12.17, $E(r, m)$ ($r \in B - \{0\}$) is the most general element of $\mathcal{A}_1(r, m)$. Therefore 12.19 supplies an estimate of the value of $f \in \mathcal{A}_1(r, m)$ at any point $x \in \Gamma \backslash G$ by means of its constant terms. This is a quantitative version of the fact that an element of $\mathcal{A}_1(r, m)$ is completely determined by its constant terms. In 12.26, this is extended to the origin.

12.20 We shall eventually let $r \to 0$. For this we need to use the form of the constant term that is also valid at the origin. Let then, as in 11.9(6),

(1)
$$\beta_{P_k, r} = (\varphi_{k,r} + \varphi_{k,-r})/2, \quad \gamma_{P_k, r} = (\varphi_{k,r} - \varphi_{k,-r})/2r.$$

For $r \neq 0$, the constant term of $E(r, m)$ at P_k can be written as

(2)
$$E(r, m)_{P_k} = \tilde{u}_k(r)\beta_{k,r} + \tilde{v}_k(r)\gamma_{k,r} = u_k(r)\varphi_{k,r} + v_k(r)\varphi_{k,-r},$$

where the first expression extends to the origin by continuity. For $r \neq 0$ we have

(3) $\qquad \tilde{u}_k(r) = u_k(r) + v_k(r), \qquad \tilde{v}_k(r) = r(u_k(r) - v_k(r)),$

(4) $\qquad 2u_k(r) = \tilde{u}_k(r) + \tilde{v}_k(r).r^{-1}, \qquad 2v_k(r) = \tilde{u}_k(r) - \tilde{v}_k(r).r^{-1}.$

These relations are clearly valid for the two expressions of the constant terms of any automorphic form of type (s, m). We let

(5) $\qquad \tilde{u}_{E(r)} = (\tilde{u}_1(r), \ldots, \tilde{u}_{l(\delta)}(r)), \qquad \tilde{v}_{E(r)} = (\tilde{v}_1(r), \ldots, \tilde{v}_{l(\delta)}(r)).$

Let $\Phi_r \colon \mathcal{A}_1(r, m) \to \mathbb{C}^{2l(\delta)}$ send $E(r, m)$ to $(\tilde{u}_{E(r)}, \tilde{v}_{E(r)})$ and let W_r be its image. We put on $\mathbb{C}^{2l(\delta)}$ the norm $|x| = \max|x_i|$. This induces a norm on W_r.

12.21 Corollary. *Let $r_n \in B - \{0\}$ and $r_n \to 0$ $(n = 1, 2, \ldots)$. For each n, let $f_n = E(r_n, m)$ be an Eisenstein series holomorphic at r_n. Assume that the set of vectors $\Phi_r(f_n)$ is bounded. Then there exists a constant $c > 0$ such that*

(1) $\qquad |r_n f_n(x)| \leqslant c \quad (x \in C, n = 1, 2, \ldots),$

(2) $\quad |r_n f_n(x)| \leqslant c.h_i^{1+d}(x) \quad \left(x \in \mathfrak{S}_{i,t} \; (i = 1, \ldots, l(\delta)), \, n = 1, 2, \ldots \right).$

From 12.20(2) we see that $|r_n.u_{f_n}|$ and $|r_n.v_{f_n}|$ are bounded as n varies, so 12.21 follows from 12.19.

12.22 Proposition. *Let $r \in \mathbb{C}^*$, and let $E = E(r)$ be an Eisenstein series that is holomorphic at r. Let $f \in \mathcal{A}_1(0, m)$. Then*

(1) $\quad 2r^2(E(r)^T, f^T) = (r.\tilde{u}_{E(r)} + \tilde{v}_{E(r)}, a_f)T^r - (r\tilde{u}_{E(r)} - \tilde{v}_{E(r)}, a_f)T^{-r}$

$\qquad\qquad\qquad + (r\tilde{u}_{E(r)} + \tilde{v}_{E(r)}, b_f)(T^r \ln T - T^r.r^{-1})$

$\qquad\qquad\qquad - (r\tilde{u}_{E(r)} - \tilde{v}_{E(r)}, b_f)(T^{-r} \ln T + T^{-r}.r^{-1}).$

Proof. Again, we may assume $E(r) = E_i(r)$ and follow the steps of the proof of 12.16. As before,

(2) $\qquad\qquad (E_i(r)^T, f^T) = M_i - \sum_{k \neq i} M_k,$

where

(3) $\qquad\qquad M_k = \int dt \, (u_k.\bar{a}_k.t^{-r-1} + u_k.\bar{b}_k.t^{-r-1} \ln t)$

and, taking 12.17(5) into account,

(4) $\qquad\qquad M_i = \int_0^T dt \, (u_i.\bar{a}_i.t^{r-1} + v_i.\bar{b}_i.t^{r-1} \ln t)$

$\qquad\qquad\qquad - \int_T^\infty dt \, (v_i.\bar{a}_i.t^{-r-1} + v_i\bar{b}_i t^{-r-1} \ln t).$

To derive (1), we substitute the expression for u_i, v_i in terms of $\tilde{u}_i$, $\tilde{v}_i$ given by 12.17(4), note that

(5) $$t^{r-1} \ln t = \frac{d}{dt}\left(\frac{t^r}{r} \ln t - \frac{t^r}{r^2}\right),$$

integrate, use (2), and multiply both sides by $2r^2$.

12.23 Corollary. *Let r_n and f_n be as in 12.21. Assume that*

$$\tilde{u}_{f_n} \to u \quad and \quad \tilde{v}_{f_n} \to v.$$

Then

(1) $$(u, b_f) = (v, a_f).$$

Proof. We consider the Maass–Selberg relations 12.22 for f_n and f and let $r_n \to 0$. As usual, $(\rho_n^T, f^T) = (f_n, f^T)$. Since f^T is fast decreasing on $\mathfrak{S}_{i,T}$ ($i = 1, \ldots, l$), it follows from 12.21 that $r_n(f_n^T, f^T)$ is bounded independently of n. Therefore, the left-hand side of 12.22(1) tends to zero.

The first two terms on the right tend to (v, a_f). The limit of

$$(r.\tilde{u}_{f_n}, b_f)(T^r \ln T - T^{-r} \ln T - T^r.r^{-1} - T^{-r}.r^{-1})$$

is $-2(u, b_f)$. Finally, using

(2) $$\lim_{r \to 0}(T^r - T^{-r}).r^{-1} = 2 \ln T,$$

we see that the limit of the remaining sum is zero, whence the corollary.

Remark. We have given three forms of the Maass–Selberg relations, each adapted to the immediate goal at the moment. We already pointed out that 12.10 is a special case of 12.16. The reader who would prefer to have just one formula, which would specialize to 12.16 and 12.23, may derive it simply by using the second form of the constant term (i.e., $\tilde{u}_E$, $\tilde{v}_E$, $\tilde{u}_f$, $\tilde{v}_f$) in proving 12.16. This essentially doubles the number of summands on the right-hand side, but otherwise brings no essential change. We leave it as an exercise.

12.24 Theorem. *The space $\mathcal{A}_1(0, m)$ has dimension $l(\delta)$.*

Proof (first part). As before, B is a compact neighborhood of the origin in $\mathbb{C}$, in which all Eisenstein series are holomorphic and W_r ($r \in B - \{0\}$) is the subspace of $\mathbb{C}^{2l(\delta)}$ spanned by the vectors $(\tilde{u}_E, \tilde{v}_E)$, where E runs through all Eisenstein series $E(r, m)$ or, equivalently (12.17), through all elements in $\mathcal{A}_1(s, m)$. The subspace W_r is $l(\delta)$-dimensional. Because the Grassmannian of $l(\delta)$-dimensional subspaces of $\mathbb{C}^{2l(\delta)}$ is compact, we can find a sequence $r_n \to 0$, $r_n \neq 0$ ($n = 1, 2, \ldots$)

such that the subspaces W_{r_n} converge to some $l(\delta)$-dimensional subspace, say W_0. It follows from 12.23 that for any $(u, v) \in W_0$ and any $f \in \mathcal{A}_1(0, m)$ we have

(1) $$(u, \tilde{v}_f) = (v, \tilde{u}_f).$$

Therefore the vectors $(\tilde{u}_f, \tilde{v}_f)$ $(f \in \mathcal{A}_1(0, m))$ are orthogonal to the $l(\delta)$-dimensional subspace spanned by the (v, u) $((u, v) \in W_0)$; hence dim $\mathcal{A}_1(0, m) \leqslant l(\delta)$. To establish the reverse inequality it suffices to show that, given $(u, v) \in W_0$, there exists $f \in \mathcal{A}_1(0, m)$ such that $\Phi_0(f) = (u, v)$. We shall more precisely produce such an f as a locally uniform limit of suitable Eisenstein series. This, however, requires some preparation. The proof will be given in 12.27 (see also 12.29).

As usual, $C(\Gamma \backslash G)$ is endowed with the compact open topology defined by the seminorms $\nu_C(f) = \sup_{x \in C} |f(x)|$ $(C \in \Gamma \backslash G$ compact$)$, and $C^\infty(\Gamma \backslash G)$ is endowed with the topology of uniform convergence on compact subsets of an element f and all of its derivatives $D.f$ $(D \in \mathcal{U}(\mathfrak{g}))$. A linear map A of $C(\Gamma \backslash G)$ (resp. $C^\infty(\Gamma \backslash G)$) into itself is said to be *compact* if any bounded sequence of functions $\{f_n\}$ contains a subsequence $\{f_{n_i}\}$ such that $\{A.f_{n_i}\}$ is convergent.

12.25 Lemma. *Let $\alpha \in C_c(G)$ (resp. $C_c^\infty(G)$). Then the convolution $*\alpha$ is a compact operator in $C(\Gamma \backslash G)$ (resp. $C^\infty(\Gamma \backslash G)$).*

This lemma is contained in [31, Lemma 59], and the proof is given here for the sake of completeness. The lemma and its proof are valid more generally if G is a unimodular locally compact group that is countable at infinity (resp. Lie group) and Γ is any discrete subgroup.

Proof. We first consider the continuous case. Let $\{f_n\}$ $(n = 1, \dots)$ be a bounded sequence in $C(\Gamma \backslash G)$ – that is, a sequence of elements on which each seminorm ν_C $(C$ compact in $\Gamma \backslash G)$ is bounded. We must show the existence of a subsequence converging locally uniformly. First, a general remark: we claim that, given a compact subset C of $\Gamma \backslash G$, there exists a constant $\delta(C)$ such that

$$|(\varphi * \alpha)(x)| \leqslant \delta(C) \|\varphi\|_{L_1} \quad \text{for } x \in C \text{ and any } \varphi \in L^1(\Gamma \backslash G).$$

Let D_o be a compact symmetric neighborhood of 1 in G containing the support of α. Fix a compact subset $D \subset G$ mapping onto C under the canonical projection $\pi: G \to \Gamma \backslash G$. If $x \in D$, then $\alpha(y^{-1}.x) \neq 0$ implies $y \in C.D_o$. Let Ω be a relatively compact open neighborhood of $C.D_o$. Then, for $x \in C$,

$$(\varphi * \alpha)(x) = \int_\Omega \varphi(y)\alpha(y^{-1}.x) \, dy.$$

If $\gamma \in \Gamma$, then $\alpha(y^{-1}\gamma x) \neq 0$ implies $\gamma \in yD_o.D^{-1}$ and hence $\gamma \in \Omega D_o.D^{-1}$, and

$$\sum{}^{\prime} |\alpha(y^{-1}\gamma x)| \leqslant \|\alpha\|_\infty \, \mathrm{Card}(\Gamma \cap \Omega.D_o.D^{-1}).$$

We let $\delta(C)$ be the right-hand side. Then

$$|(\varphi * \alpha)(x)| \leqslant \int_{\Gamma\Omega} dy.|\varphi(y)|.|\alpha(y^{-1}x)| = \int_{\Gamma\backslash\Gamma\Omega} |\varphi(y)| \sum_\gamma |\alpha(y^{-1}\gamma x)| \, dy,$$

(1) $$|(\varphi * \alpha)(x)| \leqslant \delta(C) \int_{\Gamma\backslash\Gamma\Omega} |\varphi(y)| \, dy \leqslant \delta(C)\|\varphi\|_{L_1},$$

which proves our assertion.

Let $\{B_k\}$ $(k = 1, 2, \ldots)$ be an increasing sequence of compact subsets of $\Gamma\backslash G$ such that B_k is in the interior of B_{k+1} and the union of the B_k is $\Gamma\backslash G$. Fix a continuous function β_k equal to one on B_k and to zero outside B_{k+1}. Given k, the assumption implies that the sequence $\|\beta_k f_n\|_{L_1}$ is bounded and has, therefore, a convergent subsequence. Using the diagonal process, we can then find a subsequence $\{f_{n_i}\}$ $(n_i \to \infty)$ such that, for every k, the sequence $\|\beta_k f_{n_i}\|_{L_1}$ has a limit. Changing the numbering, we may assume that our sequence $\{f_n\}$ itself has this property.

As before, let C be as before a compact subset of $\Gamma\backslash G$. We can find k such that

$$(\beta_k f_n * \alpha)(x) = (f_n * \alpha)(x) \quad (x \in C, \, n = 1, 2, \ldots).$$

By (1),

$$|(f_n * \alpha)(x)| = |(\beta_k f_n * \alpha)(x)| \leqslant \delta(C)\|\beta_k f_n\|_{L_1}.$$

This means that $f_n * \alpha$ converges uniformly and absolutely on C.

If now $\alpha \in C_c^\infty(G)$ then we have $A(f_n * \alpha) = f * A.\alpha$ for any $A \in \mathcal{U}(\mathfrak{g})$, so that $f_n * \alpha$ converges in the C^∞-topology.

12.26 Proposition. *Let B and d be as in 12.18, and choose a constant*

$$\lambda \geqslant 2(1 + d).$$

Then there exists a constant $C > 0$ such that

(1) $$|f(x)| \leqslant C.|\Phi_r(f)|.h_{P_i}^\lambda(x) \quad (x \in \mathfrak{S}_{i,t}, \, i = 1, \ldots, l)$$

for every $r \in B$ and $f \in \mathcal{A}_1(r, m)$.

Proof. We note first that, given $f \in \mathcal{A}_1(r, m)$, there exists a constant $d(f)$ such that

(2) $$|f(x)| \leqslant d(f).|\Phi_r(f)|.h_{P_i}^\lambda(x) \quad (x \in \mathfrak{S}_{i,t}, \, i = 1, \ldots, l).$$

In fact, $f = (f - f_{P_i}) + f_{P_i}$ and on $\mathfrak{S}_{i,t}$ the first term is bounded, even fast decreasing, while $f_{P_i}(x)$ is bounded by $|\Phi_r(f)|h_{P_i}^\lambda$, as follows from

$$f_{P_i} = \tilde{u}_i(f)\beta_{i,s} + \tilde{v}_i(f)\gamma_{i,s} = \varphi_{P_i,m} h_{P_i}\left(\tilde{u}_i(f)\cosh h^r_{P_i} + \tilde{v}_i(f).(\sinh h^r_{P_i}).r^{-1}\right)$$

(see 12.20(1)). The constants $d(f)$ for which (2) holds have a strictly positive minimum, which we denote $c(f)$. The proposition amounts to stating that $c(f)$ is bounded on the union of the $A_1(r, m)$ $(r \in B)$. The proof is by contradiction.

Assume it is not so. Then we can find $r_n \in B$ $(n = 1, 2\ldots)$ and $f_n \in A_1(r, m)$ such that $c(f_n) \to \infty$. Moreover, we may assume that $|\Phi_r(f_n)| = 1$ (if not, replace f_n by $|\Phi(f_n)|^{-1}.f_n$, which does not change $c(f_n)$) and, passing to a subsequence if necessary, that $r_n \to r_0 \in B$. In view of the definition of $c(f_n)$, there exists $i \in [1, l]$ and $x \in \mathfrak{S}_{i,t}$ such that

(3) $$|f_n(x)| h_{P_i}(x)^\lambda \geqslant c(f_n)/2$$

(i may depend on n but, again passing to a subsequence, we may assume this to be true for all n). It will be convenient to introduce the seminorm

$$\nu_\lambda(f) = \sup|f(x)|.h_{P_i}(x)^{-\lambda} \quad (x \in \mathfrak{S}_{i,t}, \ i = 1, \ldots, l),$$

which is, of course, finite only on those f with growth in all $\mathfrak{S}_{i,t}$ essentially bounded by $h_P(x)^\lambda$ – but this is the case for all f under consideration in this proof. Let $g_n = c(f_n)^{-1}.f_n$. Then, taking into account the definition of $c(f_n)$ and (3), we have

(4) $$1/2 \leqslant \nu_\lambda(g_n) \leqslant 1 \quad (n \geqslant 1).$$

The g_n are all Eisenstein functions, so

$$g_n * \alpha = \pi_\alpha(r_n).g_n \quad (n \geqslant 1),$$

which can be written as

(5) $$g_n = g'_n * \alpha, \quad \text{where } g'_n = \pi_\alpha(r_n)^{-1} g_n.$$

Recall that, by the definition of compatibility, $|\pi_\alpha(r)|^{-1} \leqslant 2$ on B. Therefore

(6) $$\nu_\lambda(g'_n) \leqslant 2 \quad (n \geqslant 1).$$

As a consequence, on any compact subset C there exists $\delta(C)$ such that

$$\sup_{x \in C}|g'_n(x)| \leqslant \delta(C)$$

for all n. By 12.25 there is a subsequence $g'_n * \alpha$ that converges locally uniformly in the C^∞ topology. Let g_0 be the limit of such a subsequence. We want to prove that $g_0 = 0$. After renumbering, we may assume that $g_n \to g_0$. We show that $g_0 \in A_1(r_0, m)$. First, passing to the limit in the inequality $\nu_\lambda(g_n) \leqslant 1$, we obtain $\nu_\lambda(g_0) \leqslant 1$ and hence g_0 is of moderate growth; it is clearly of right K-type m. Since the convergence is in the C^∞ topology, $Cg_n \to Cg_0$. But $Cg_n = ((r_n^2 - 1)/2)g_n$, hence $\mathcal{C}g_0 = ((r_0^2 - 1)/2)g_0$, so g_0 is an automorphic form. Let now φ be a cusp form. We claim that

$$\int_{\Gamma\backslash G} \varphi(x).\overline{g_n(x)}\,dx \to \int_{\Gamma\backslash G} \varphi(x).\overline{g_0(x)}\,dx.$$

On any compact set, this is true by uniform convergence. The function φ is rapidly decreasing on $\mathfrak{S}_{i,t}$; therefore, given $\varepsilon > 0$, there exists T such that

$$|\varphi(x).h_{P_i}^\lambda(x)| \leqslant \varepsilon \quad (x \in \mathfrak{S}_{i,T},\ i = 1, \dots, l)$$

and also, for all $n \geqslant 1$, $\int_{\mathfrak{S}_{i,T}} |\varphi(x).g_n(x)|\,dx \leqslant \varepsilon \operatorname{vol}\mathfrak{S}_{i,T}$. Because the complement of the union of the $\mathfrak{S}_{i,T}$ is compact, our assertion follows, and this shows that g_0 is orthogonal to cusp forms and so $g_0 \in \mathcal{A}_1(r_0, m)$. For any cuspidal subgroup P, the constant term $g_{n,P} \to g_{0,P}$. But we have

$$(7) \qquad |\Phi_{r_n}(g_n)| = c(f_n)^{-1}.|\Phi_{r_n}(f_n)| = c(f_n)^{-1} \to 0;$$

therefore g_0 is a cusp form in $\mathcal{A}_1(r_0, m)$ and hence is zero. For $D \in \mathcal{U}(\mathfrak{g})$,

$$Dg_n = D(g'_n * \alpha) = g'_n * D\alpha,$$

whence, in view of (6), the existence of a constant $c(D)$ such that

$$(8) \qquad \nu_\lambda(Dg_n) \leqslant c(D) \quad \text{(for all } n \geqslant 1).$$

As in 11.16, we now see that if we apply 7.5 to $g_n - g_{n,P_i}$ then the implied constants may be chosen *independently* of n; that is, given $N \in \mathbb{N}$, there exists $c_N > 0$ such that

$$(9) \quad |g_n(x) - g_{n,P_i}(x)| \leqslant c_N.h_{P_i}^{-N}(x) \quad \left(x \in \mathfrak{S}_{i,t}\ (i = l, \dots, l),\ n \geqslant 1\right).$$

On the other hand, (7) implies that

$$|g_{n,P_i}(x).h_{P_i}^{-\lambda}(x)| \leqslant c(f_n)^{-1} \quad \left(x \in \mathfrak{S}_{i,t}\ (i = 1, \dots, l),\ n \geqslant 1\right).$$

We can find $T > 0$ and $N \in \mathbb{N}$ such that $c_N h_{P_i}(x)^{-N-\lambda} \leqslant 1/4$ for $x \in \mathfrak{S}_{i,T}$ $(i = 1, \dots, l)$. Then

$$(10) \quad |g_n(x)h_{P_i}(x)^{-\lambda}| \leqslant h_{P_i}^{-\lambda}(x)|g_n(x) - g_{n,P_i}(x)| + |g_{n,P_i}(x)| \leqslant 1/4 + c(f_n)^{-1}$$

for $x \in \mathfrak{S}_{i,T}$ $(i = 1, \dots, l)$ and $n \geqslant 1$. For n big enough, we have

$$(11) \qquad |g_n(x)h_{P_i}(x)^{-\lambda}| < 1/2 \quad \left(x \in \mathfrak{S}_{i,T}\ (i = 1, \dots, l),\ n \geqslant 1\right).$$

This shows that a point x in some $\mathfrak{S}_{i,t}$ at which

$$|g_n(x)|.h_{P_i}^{-\lambda}(x) \geqslant 1/2,$$

which should exist in view of (4), must belong to the complement of the union of the smaller Siegel sets $\mathfrak{S}_{i,T}$. But this complement is compact, which contradicts that $g_n \to 0$ uniformly on any compact set. This contradiction proves the proposition.

Remark. The proof of 12.26 is in substantial part extracted from that of Theorem 6 in [31]. From 12.19 one sees easily that the point r_0 must be the origin, so that the crux of the matter is at the origin.

We note that the proof of 12.19 is valid around any point $\neq 0$ at which all Eisenstein series are holomorphic. Therefore we have an estimate of $f \in \mathcal{A}_1(r, m)$ in terms of its constant terms at all points at which the E_j are holomorphic. Using 12.19, it would not be too hard to extend this to the other points. We leave this as an exercise.

12.27 Proof of 12.24 (second part). Let $(u, v) \in W_0$. There exists a sequence of elements $r_n \in B - \{0\}$ tending to the origin and vectors $(u_n, v_n) \in W_{r_n}$ such that $u_n \to u$ and $v_n \to v$. Let $f_n \in \mathcal{A}_1(r_n, m)$ be the unique element such that

$$\Phi_{r_n}(f_n) = (u_n, v_n).$$

We claim that f_n tends locally uniformly to an element $f_0 \in \mathcal{A}_1(0, m)$ such that $\Phi_0(f_0) = (u, v)$. This will then show that $\dim \mathcal{A}_1(0, m) \geq l(\delta)$ and conclude the proof of 12.24.

As before, we can write

$$(1) \qquad f_n = f_n' * \alpha, \quad \text{where } f_n' = \pi_\alpha(r_n)^{-1} . f_n \quad (n \geq 1).$$

By construction, $\Phi_{r_n}(f_n) = (u_n, v_n)$ is bounded independently of n. Since

$$|\pi_\alpha(r_n)^{-1}| \leq 2,$$

this is also true of $\Phi_{r_n}(f_n')$ $(n = 1, \ldots)$. By 12.26, there exists then a constant $q > 0$ such that

$$(2) \qquad |f_n'(x)| \leq q . h_{P_i}^\lambda(x) \quad \left(x \in \mathfrak{S}_{i,t} \ (i = 1, \ldots, l), \ n \geq 1\right).$$

In particular, for any compact subset C of $\Gamma \backslash G$, the set $\{\nu_C(f_n')\}_{n \geq 1}$ is bounded. By 12.25, there is then a subsequence $\{n_i\}$ such that f_{n_i} converges locally uniformly on every compact set. Let f_0 be that limit. Then, exactly as in 12.26 (for g_0) we see that $f_0 \in \mathcal{A}_1(0, m)$ and $\Phi_0(f_0) = (u, v)$.

Thus, any infinite subsequence of the $\{f_n\}$ contains a subsequence that converges locally uniformly to an element $g \in \mathcal{A}_1(0, m)$ such that $\Phi_0(g) = (u, v)$. This implies that $g = f_0$; therefore, the sequence $\{f_n\}$ itself converges locally uniformly to f_0.

12.28 Remark on holomorphic automorphic forms. Let $m \in \mathbb{N}$ and $m \geqslant 3$. We come back to 10.8 and consider the series (4), which we now write as

$$Q(P, m) = \sum_{\gamma \in \Gamma_P \backslash \Gamma} \mu(\gamma, z)^{-m}.$$

This series is a holomorphic automorphic from on X of weight m (the convergence follows from 10.4). Its lifting $\tilde{Q}(P, m)$ (5.14) is an automorphic form on $SL_2(\mathbb{R})$ of right K-type m that is an eigenfunction of C with eigenvalue $((m-1)^2 - 1)/2$ (5.15). The constant terms of $Q(P, m)$ are genuine constants (7.2), and 10.4(v) implies that the constant terms at the cusps not Γ-equivalent to P are zero so that, again, the Eisenstein series $Q(P_i, m)$ ($i \leqslant l(\delta)$) are linearly independent, and 12.17 shows that the $\tilde{Q}(P_i, m)$ span $\mathcal{A}_1(m - 1, m)$. Consequently, the $Q(P_i, m)$ span the orthogonal complement of the space of holomorphic cusp forms of weight m in the space of holomorphic automorphic forms of weight m (as is well known). Theorems 12.23 and 12.25 in [53] give the dimensions of those spaces, and also imply that $\dim \mathcal{A}_1(m - 1, m) = l(\delta)$.

12.29 Remark (added in proofs). Sections 12.25 and 12.26 were mainly a preparation for the proof of

(1) $$\dim \mathcal{A}_1(0, m) \geqslant l(\delta).$$

They are of independent interest, but could be avoided here. Recently, J-L. Waldspurger communicated to me a more direct proof. In fact, it is basically the argument in the first part of 12.17, but where $c(E)$ is replaced by minus the order of the zero of $E(r, m)$ at the origin. To see this, first note that the proof at the end of 11.9, leading to 11.10 and 11.11, remains valid, mutatis mutandis, for orders of zeroes, rather than of poles, $m(c)$ being replaced by minus the order of a zero. More precisely, if $E(r, m)$ is holomorphic in r around c, then $E(r, m)$ has a zero of order $\geqslant j$ at c if and only if its constant terms do so; moreover, if it does, then

$$\lim_{r \to c} (r - c)^{-j} . E(r, m)$$

is an element of $\mathcal{A}_1(c, m)$.

Choose a neighborhood U of the origin in which all Eisenstein series are holomorphic and fix $z \in U$, $z \neq 0$. We want to establish an injective linear map of $\mathcal{E}(r, m)$ into $\mathcal{A}_1(0, m)$. Let d be the maximum of the orders of the zeroes of the $E(r, m)$ at 0. For $0 \leqslant j \leqslant d$, let A_j be the subspace of $\mathcal{E}(z, m)$ spanned by the Eisenstein series that have a zero of order at least j at the origin. On A_j, we consider the operator

(2) $$\pi_j: E(z, m) \to \lim_{r \to 0} r^{-j}.E(r, m).$$

By what was just said, its image C_j belongs to $\mathcal{A}_1(0, m)$. Its kernel is A_{j+1}, hence π_j is an isomorphism of a supplement B_j of A_{j+1} in A_j onto C_j. It is then seen as in 12.17 that the spaces C_j are linearly independent, and (1) follows.

Spectral decomposition and representations

13 Spectral decomposition of $L^2(\Gamma\backslash G)_m$ with respect to C

13.0 From now on, H and $^\circ H$ will stand for $L^2(\Gamma\backslash G)$ and $^\circ L^2(\Gamma\backslash G)$, respectively, and H^∞ is as in 11.12. As before,

$$(\varphi, \psi) = \int_{\Gamma\backslash G} \varphi(g).\overline{\psi(g)}\, dg \quad (\varphi, \psi \in H)$$

denotes the scalar product on H.

For $m \in \mathbb{Z}$, set

$$H_m = \{ f \in H \mid f(xk) = f(x)\chi_m(k)\ (x \in G,\, k \in K) \}.$$

If V is a closed right K-stable subspace of H, then $V_m = V \cap H_m$.

The notation of 10.12 and 12.0 is assumed.

13.1 Lemma. *Let $\alpha, \alpha_n \in C_c^\infty(G)$ $(n = 1, 2, \ldots)$, assume that the α_n form a Dirac sequence (2.4), and let $f \in H$.*

(i) *The convolution $*\alpha$ by α is a bounded operator on H, with adjoint $*\alpha^*$.*
(ii) *The function $x \mapsto r_x f$ mapping G into H is continuous.*
(iii) *The sequence $f * \alpha_n \to f$ in H as $n \to \infty$. In particular, H^∞ is dense in H.*

Proof. (i) Let C be a compact symmetric neighborhood of 1 containing the support of α and hence also that of α^*. We have

$$\|f * \alpha\|^2 = \int_{\Gamma\backslash G} |f * \alpha(x)|^2\, dx \leqslant \int_{\Gamma\backslash G} dx \left(\int_C |f(xy)\alpha(y^{-1})|\, dy \right)^2.$$

By the Schwarz inequality,

$$\left(\int_C |f(xy)\alpha(y^{-1})|\, dy \right)^2 \leqslant \left(\int_C |f(xy)|^2\, dy \right) \cdot \left(\int_C |\alpha(y^{-1})|^2\, dy \right);$$

hence

$$\|f * \alpha\|^2 \leqslant \int_{\Gamma\backslash G} dx \left(\int_C |f(xy)|^2\, dy \right) \cdot \left(\int_C |\alpha(y^{-1})|^2\, dy \right)$$

$$\leqslant \left(\int_G |\alpha(y^{-1})|^2\, dy \right) \cdot \int_C dy\, \|r_y f\|^2 \leqslant D\|f\|^2,$$

which shows that $*\alpha$ is a bounded operator. The second part of (i) follows from an elementary computation, which is carried out in a slightly more general framework in 14.2.

(ii) We have to show that, given $\varepsilon > 0$ and $x \in G$, there exists a neighborhood U of x such that $\|r_z f - r_x f\| < \varepsilon$ for all $z \in U$.

Assume first that $f \in C_c(\Gamma \backslash G)$. Then, by uniform convergence, given $c > 0$ we can find U such that $|r_z f(y) - r_x f(y)| < c$ for all $z \in U$ and $y \in \Gamma \backslash G$, whence our assertion in this case. Because the space $C_c(\Gamma \backslash G)$ is dense in $L^2(\Gamma \backslash G)$, in the L^2-norm, we can choose a sequence $f_m \in C_c(\Gamma \backslash G)$ such that $f_m \to f$ in $L^2(\Gamma \backslash G)$. For $z \in G$ and $m \in \mathbb{N}$ we have

$$\|r_z f - r_x f\| \leqslant \|r_z f - r_z f_m\| + \|r_z f_m - r_x f_m\| + \|r_x f_m - r_x f\|.$$

There exists $m \in \mathbb{N}$ such that $\|f_m - f\| < \varepsilon/3$, whence

$$\|r_z f - r_z f_m\| = \|r_x f_m - r_x f\| = \|f_m - f\| < \varepsilon/3.$$

Then, by what has already been proved, given m there exists a neighborhood U of x such that

$$\|r_z f_m - r_x f_m\| < \varepsilon/3 \quad (z \in U),$$

and (ii) follows.

(iii) Let C be a compact neighborhood of the identity containing the support of the α_n. Given $\varepsilon > 0$, there exists by (ii) a neighborhood U of 1 such that $\|r_y f - f\| < \varepsilon$ for $y \in U$. Clearly,

$$f * \alpha_n(x) - f(x) = \int_C \alpha_n(y^{-1})(f(xy) - f(x)) \, dy;$$

hence

$$\|f * \alpha_n - f\| \leqslant \int_C \alpha_n(y^{-1})\|r_y f - f\| \, dy \leqslant \varepsilon \int_G \alpha_n(y^{-1}) \, dy = \varepsilon.$$

Since the elements $f * \alpha_n$ all belong to H^∞, this concludes the proof of (iii).

Remark. The previous lemma suffices for our needs in Section 13. In Section 14, it will be put in a more general and more natural framework.

13.2 The essential self-adjointness of C.   In order to speak of spectral decomposition with respect to C, we must know that C is essentially self-adjoint. We reduce the proof here to the main criterion for self-adjointness. We assume familiarity with some generalities on unbounded operators (see e.g. [46, VIII.1, VIII.2]). We recall that an unbounded operator S on some Hilbert space comes equipped with a domain of definition $D(S)$. If T is another operator, if $D(S) \subset D(T)$, and if T coincides with S on $D(S)$, then T is said to be an *extension* of S, denoted $S \subset T$.

We now view C as an unbounded operator on H with domain of definition H^∞. It is *symmetric* by 11.12(4) and $D(C)$ is dense by (13.1), so the adjoint C^* of C is defined and $C^* \supset C$. An adjoint is always closed, so C has a minimal closure $\bar{C}$, with graph the closure of the graph of C. We claim that $\bar{C}$ is self-adjoint – in other words, that C is *essentially self-adjoint*. By the basic criterion for self-adjointness [46, Thm. VIII.3], it suffices to prove that $\pm i$ are not eigenvalues of C^* on $D(C^*)$. This is equivalent to C^* not having nonreal eigenvalues [46, Thm. X.1], and that is in fact what we prove.

By 2.3(iii) we know that

(1) $(C\varphi) * \alpha = C(\varphi * \alpha)$ $(\varphi \in H^\infty,\ \alpha \in C_c^\infty(G))$.

We claim that we have similarly

(2) $(C^*\varphi) * \alpha = C^*(\varphi * \alpha)$ $(\varphi \in D(C^*),\ \alpha \in C_c^\infty(G))$.

To see this, it suffices to prove

(3) $((C^*\varphi) * \alpha, \mu) = (C^*(\varphi * \alpha), \mu)$ for all $\mu \in C_c^\infty(\Gamma\backslash G)$.

Let $\mu \in C_c^\infty(\Gamma\backslash G)$. By definition of the adjoint, and since $\mu \in D(C)$, we have, also using 13.1(i),

(4) $(C^*(\varphi * \alpha), \mu) = (\varphi * \alpha, C\mu) = (\varphi, C\mu * \alpha^*)$.

Similarly, and taking (1) into account,

(5) $(C^*(\varphi) * \alpha, \mu) = (C^*(\varphi), \mu * \alpha^*) = (\varphi, C(\mu * \alpha^*)) = (\varphi, C\mu * \alpha^*)$,

so (2) follows from (4) and (5).

Assume now that $C^*\varphi = \lambda\varphi$ for some $\varphi \in D(C^*)$ ($\varphi \neq 0$) and $\lambda \in \mathbb{C}$. We obtain from (2) that $C^*(\varphi * \alpha) = \lambda(\varphi * \alpha)$. Moreover, by using a Dirac sequence, we can arrange that $\varphi * \alpha \neq 0$ (13.1). But $\varphi * \alpha \in D(C)$, hence $C(\varphi * \alpha) = \lambda(\varphi * \alpha)$ and λ is real.

It is a general fact that an operator and its closure (assumed to exist) have the same adjoint. Thus we have $\bar{C} = C^*$.

13.3 Remarks.   (i) We have chosen H^∞ as domain of definition of C. This is the most natural from the point of view of representation theory (see §14). We could also have started from $C_c^\infty(\Gamma\backslash G)$, but the adjoint is the same. More precisely, let C_1 be the unbounded operator defined by the Casimir operator on $C_c^\infty(\Gamma\backslash G)$. It is symmetric and $C_1 \subset C \subset C^*$, hence also $\bar{C}_1 \subset \bar{C} = C^*$. We claim that

(1) $C_1^* = C^*$

(hence also $\bar{C}_1 = \bar{C}$). The relation 11.12(4) shows that $H^\infty \subset D(\bar{C}_1)$. Then the proof of 13.2 remains valid if C is replaced by $\bar{C}_1$, showing that $\bar{C}_1 = C_1^*$ and therefore $C_1^* \subset C^*$, whereas the inclusion $C_1 \subset C$ also yields $C^* \subset C_1^*$.

(ii) The previous argument proves essential self-adjointness in more general cases (see §14.9).

13.4 Theorem. *The spectrum of C in $°H_m$ is discrete, with finite multiplicities. The space $°H_m$ has a Hilbert space basis consisting of countably many cusp forms that are eigenfunctions of C. In particular, the cusp forms are dense in $°H_m$.*

Proof. The basic fact to be used is that $*\alpha$ is a compact operator on $°H$ for $\alpha \in C_c^\infty(G)$ (9.3). Let $\{\alpha_n\}$ ($n \in \mathbb{N}$) be a Dirac sequence and let us assume, as we may, that the α_n are symmetric. Then the operators $*\alpha_n$ are self-adjoint (13.1(i)). For each n, the space $°H_m$ is a countable direct sum of eigenspaces of $*\alpha_n$, and the eigenspaces for nonzero eigenvalues are finite dimensional by the Riesz–Schauder theorem (see [61, X.5, p. 283] or [46, VI.5, p. 203]). These finite dimensional eigenspaces for the α_n span $°H_m$; otherwise, there would exist $f \in °H_m$ ($f \neq 0$) such that $f * \alpha_n = 0$ for all n, which contradicts 13.1(iii).

Let E be one such eigenspace. Since C commutes with $(*\alpha_n)$, it leaves E stable. Then E consists of K-finite and C-finite cuspidal functions, hence of cusp forms. The restriction of C to E is self-adjoint and so can be diagonalized. This shows that $°H_m$ has a basis consisting of eigenfunctions of C, which are then cusp forms, all of K-type m. Therefore the eigenvalue of C determines the type of the cusp form. Those of a given $C - K$-type form a finite dimensional space by 8.5, from which it follows that the spectrum of C in $°H_m$ has finite multiplicities.

There remains then to investigate the orthogonal complement V_m of $°H_m$ in H_m.

13.5 Lemma. *The orthogonal complement V_m of $°H_m$ in H_m contains the residues of Eisenstein series at the poles on $(0, 1]$ and the E-series $pE_{u,m}$, $u \in C_c^\infty(A)$ (see 12.2), where (P, A) runs through the cuspidal p-pairs. It is spanned by those series when P runs through $P_1, \ldots, P_{l(\delta)}$.*

Proof. Let ψ be an automorphic form of K-type m. Then

$$(1) \qquad (\psi, {}_pE_{u,m})_{\Gamma \backslash G} = (\psi_P, u)_{N \backslash G} = \int_A h_P(a)^{-2} . \psi_P(a) . \bar{u}(a) \, da$$

(see 12.3 and the proof of 12.6). Therefore, $_pE_{u,m}$ is orthogonal to all cusp forms. Since those span $°H_m$ by 13.4, $_pE_{u,m} \in V_m$.

Conversely, if ψ is orthogonal to all $_pE_{u,m}$ for a given P, then by (1) its constant term is orthogonal to all $u \in C_c^\infty(A)$ with respect to the measure $h_P(a)^{-2} \, da$ and hence is identically zero. If this holds for $P = P_1, \ldots, P_{l(\delta)}$, then the constant terms of ψ at the P_i are zero and ψ is a cusp form (7.8), which proves the last assertion. Finally, the residual Eisenstein series at the points of $(0, 1]$ are square integrable by 12.11 and orthogonal to cusp forms by 11.11.

13.6 Notation. As before, we fix $m \in \mathbb{Z}$ and write $E_j(s)$ for $E_j(s, m)$. Let J_m be the set of $z \in (0, 1]$ at which at least one $E_j(s)$ has a pole or, equivalently by 12.12, at which one of the $c_{j,j}(s)$ has a pole. Those poles are simple and are the only possible poles in the right half-plane $\mathcal{R}s \geqslant 0$; moreover, J_m is finite (11.13).

For $z \in (0, 1]$, let $I_m(z)$ be the set of j for which $E_j(s)$ has a pole at z. The $I_m(z) \times I_m(z)$ matrix

$$C'(z) = (c'_{j,k}(z)) \quad (j, k \in I_m(z))$$

is a Hermitian matrix with nonzero diagonal terms, containing all nonzero residues at z of the $c_{j,k}$ $(1 \leqslant j, k \leqslant l)$. We denote by $F_j(z)$ the unit vector $E'_j(z).\|E'_j(z)\|^{-1}$ spanned by $E'_j(z)$.

Let $u_j \in C_c^\infty(A_j)$ and $v_k \in C_c^\infty(A_k)$. Let $\psi_j = P_j E_{u_j, m}$ and $\eta_k = P_k E_{v_k, m}$ $(1 \leqslant j, k \leqslant l(\delta))$.

13.7 Theorem. *We keep the notation of 13.6. Then*

$$(1) \qquad (\psi_j, \eta_k) = (2\pi)^{-1} \sum_{n \leqslant l(\delta)} \int_0^\infty (\psi_j, E_n(i\tau)).(E_n(i\tau), \eta_k)\, d\tau$$

$$+ \sum_{z \in J_m} (\psi_j, F_j(z))(F_k(z), \eta_k)(F_j(z), F_k(z)).$$

Proof. By 12.6, for $\sigma > 1$ we have

$$(\psi_j, \eta_k) = \int_{\sigma + i\mathbb{R}} \hat{u}_j(s)\{\delta_{j,k}\overline{\hat{v}_k(-\bar{s})} + c_{j,k}(s)\overline{\hat{v}_k(\bar{s})}\}\, d\tau.$$

We claim that we can shift the integral to the imaginary axis, provided we add the residues at J_m. Given $T > 0$, we integrate the integrand of the right-hand side on the boundary of a rectangle $\mathcal{R}s \in [0, \sigma]$ and $|\mathcal{I}s| \leqslant T$. The $c_{j,k}$ are bounded in the region $\mathcal{R}s \in [0, \sigma], |\mathcal{I}s| \geqslant T$ by 11.13. The functions $\hat{u}_j, \hat{v}_k$ are fast decreasing there. Therefore the integrals over the horizontal segments tend to 0 as $T \to \infty$, so

$$(2) \qquad (\psi_j, \eta_k) = \int_{i\mathbb{R}} \hat{u}_j(s)\{\delta_{j,k}\overline{\hat{v}_k(s)} + c_{j,k}(s)\overline{\hat{v}_k(-s)}\}\, d\tau$$

$$+ 2\pi \sum_{z \in J_m} \hat{u}_j(z)\overline{\hat{v}_k(z)}c'_{j,k}(z).$$

By 12.6(iii) and 12.12,

$$(3) \qquad (2\pi)^{1/2}\hat{u}_j(z) = (\psi_j, F_j(z))\|E'_j(z)\|^{-2},$$

$$(4) \qquad (2\pi)^{1/2}\overline{\hat{v}_k(z)} = (F_k(z), \eta_k)\|E'_k(z)\|^{-2},$$

whence, again using 12.12,

(5) $2\pi.\hat{u}_j(z).\overline{\hat{v}_j(z)}.c'_{j,k}(z) = (\psi_j, F_j(z)).(F_k(z), \eta_k).(F_j(z), F_k(z)).$

This shows that the second terms in the right-hand sides of (1) and (2) are equal. To prove the equality of the first terms, we first show

(6) $\displaystyle\sum_{n\leqslant l(\delta)} \int_{i\mathbb{R}} (\psi_j, E_n(s))(E_n(s), \eta_k)\, d\tau$

$$= 4\pi \int_{i\mathbb{R}} \hat{u}_j(s)\{\delta_{j,k}\overline{\hat{v}_k(s)} + c_{j,k}(s)\overline{\hat{v}_k(-s)}\}\, d\tau.$$

This will follow from the functional equation.

By 12.6 and 12.7, on $i\mathbb{R}$ we have

(7) $(\psi_j, E_n(s))E_n(s) = (2\pi)^{1/2}\{\delta_{j,n}\hat{u}_j(s) + c_{j,n}(-s).\hat{u}_j(-s)E_n(s)\}.$

Summing over n, we have

$$\sum_{n\leqslant l(\delta)} (\psi_j, E_n(s))E_n(s) = (2\pi)^{1/2}\left\{\hat{u}_j(s)E_j(s) + \hat{u}_j(-s)\sum_{n\leqslant l(\delta)} c_{j,n}(-s)E_n(s)\right\}.$$

The last sum is equal to $E_j(-s)$ by the functional equation (11.9), whence

(8) $\displaystyle\sum_{n\leqslant l(\delta)} (\psi_j, E_n(s)).E_n(s) = (2\pi)^{1/2}\{\hat{u}_j(s)E_j(s) + \hat{u}_j(-s)E_j(-s)\}.$

Taking the scalar product with η_k and integrating over $i\mathbb{R}$, we obtain

(9) $\displaystyle\sum_{n\leqslant l(\delta)} \int_{i\mathbb{R}} (\psi_j, E_n(s))(E_n(s), \eta_k)\, d\tau = 2(2\pi)^{1/2}\int_{i\mathbb{R}} \hat{u}_j(s)(E_j(s), \eta_k)\, d\tau,$

which, together with 12.6, yields (6). In view of (2) and (5), in order to finish the proof of (1) it only remains to show

(10) $\displaystyle\sum_{n\leqslant l(\delta)} \int_{-\infty}^{\infty} (\psi_j, E_n(i\tau)).(E_n(i\tau), \eta_k)\, d\tau$

$$= 2\sum_{n\leqslant l(\delta)} \int_0^{\infty} (\psi_j, E_n(i\tau)).(E_n(i\tau), \eta_k)\, d\tau;$$

but this follows from (8), since the right-hand side is obviously an even function of s.

13.8 Spectral decomposition of an incomplete theta series. From 13.7 we obtain

(1) $\displaystyle\|\psi_j\|^2 = (2\pi)^{-1}\sum_{n\leqslant l(\delta)} \int_0^{\infty} |(\psi_j, E_n(i\tau))|^2\, d\tau + \sum_{z\in J_m} |(\psi_j, F_j)|^2.$

By the Schwarz inequality, this implies that, for any $n \in \{1, \dots, l(\delta)\}$,

(2) $$\left| \int_0^\infty (\psi_j, E_n(i\tau)).(\eta_k, E_n(i\tau)) \, d\tau \right| \leqslant 2\pi \|\psi_j\| \|\eta_k\|;$$

therefore, the map

(3) $$\lambda(\psi_j, n): \eta_k \mapsto (2\pi)^{-1} \int_0^\infty (\psi_j, E_n(i\tau))(\eta_k, E_n(i\tau)) \, d\tau$$

is a bounded linear form on the span of the η_k $(k = 1, \ldots, l(\delta))$ of norm $\leqslant 2\pi \|\psi_j\|$. But this space is dense in V_m (13.5), so $\lambda(\psi_j, n)$ extends to a continuous linear form on V_m. By the Riesz representation theorem ([61, §III.6] or [46, II.4]) there is a unique element of V_m, here denoted $\mathcal{E}_n(\psi_j)$, such that

(4) $$\lambda(\psi_j, n)(x) = (x, \mathcal{E}_n(\psi_j)) \quad (x \in V_m).$$

We also write

(5) $$\mathcal{E}_n(\psi_j) = (2\pi)^{-1} \int_0^\infty (\psi_j, E_n(i\tau)).E_n(i\tau) \, d\tau.$$

Let $R_{m,z}$ be the subspace of V_m spanned by the $E_j'(z)$ (recall that $E_j(s) = E_j(s, m)$ by convention) for $z \in J_m$. These subspaces are mutually orthogonal since they belong to different eigenvalues of C. Let R_m be the direct sum of the $R_{m,z}$ $(z \in J_m)$; it is a finite dimensional subspace of V_m. Choose an orthonormal basis r_q $(q \in Q)$ of R_m that is the union of orthonormal bases of the $R_{m,z}$.

In accordance with notation to be introduced later (16.5, 17.1), we also write $H_{rs,m}$ for R_m and let $H_{ct,m}$ be the orthogonal complement of $H_{rs,m}$ in V_m.

The eigenvalues of C in $H_{rs,m}$ are of the form $(\sigma^2 - 1)/2$ with $\sigma \in (0, 1]$, and so are greater than $-1/2$. On the other hand, $E_n(i\tau)$ belongs to the eigenvalue $-(\tau^2 + 1)/2 \leqslant -1/2$. It follows therefore from the spectral theorem for self-adjoint operators [46, VIII.3; 61, §XI.6] that $\mathcal{E}_n(\psi_j)$ belongs to $H_{ct,m}$. Let p_{rs} and p_{ct} be the orthogonal projections on $H_{rs,m}$ and $H_{ct,m}$, respectively. Then

(6) $$p_{ct}(\psi_j) = \sum_{n \leqslant l(\delta)} \mathcal{E}_n(\psi_j), \qquad p_{rs}(\psi_j) = \sum_{z \in J_m} (\psi_j, F_j(z)).F_j(z).$$

This also shows that $R_{m,z}$ is the full eigenspace of C for the eigenvalue $(z^2 - 1)/2$, hence also that $H_{rs,m}$ is the sum of the eigenspaces of C in V_m for the eigenvalues $(z^2 - 1)/2$ $(z \in J_m)$.

Let ψ_j, η_k be incomplete theta series $(1 \leqslant j, k \leqslant l(\delta))$; let $\psi = \sum_j \psi_j$ and $\eta = \sum_k \eta_k$. We can extend the foregoing by linearity and define $\mathcal{E}_n(\psi)$ to represent the sum of the $\lambda(\psi_j, n)$. Then

(7) $$(p_{ct}(\psi), \eta) = (2\pi)^{-1} \sum_{n \leqslant l(\delta)} \int_0^\infty (\psi, E_n(i.\tau)).(E_n(i\tau), \eta) \, d\tau;$$

this is, of course, also equal to $(p_{ct}(\psi), p_{ct}(\eta))$. The simplest way for this to be consistent with 13.8 is to have

(8) $(\mathcal{E}_j(\psi), \mathcal{E}_k(\eta)) = \delta_{j,k} \int_0^\infty (\psi, E_j(i\tau)).(E_k(i\tau), \eta) \, d\tau,$

which indeed will follow from 17.6. The proof does not use representation theory
and could be given here, but we prefer to describe it in the framework of Section
17. It is very similar, in statement and proof, to Proposition 7.1 of [36] (which
deals with V_0). This shows that the decomposition

(9) $$\psi = \sum_{n \leqslant l(\delta)} \mathcal{E}_n(\psi) + \sum_{q \in Q} (\psi, r_q).r_q$$

is an orthogonal spectral decomposition of ψ.

By 13.5, the incomplete theta series are dense in V_m. We shall also see that those
which are contained in $H_{ct,m}$ are dense there (17.2).

13.9 Remarks. (i) If $m = 0$ then $L^2(\Gamma \backslash G)_m$ is the space of right K-invariant
functions in $L^2(\Gamma \backslash G)$, that is, $L^2(\Gamma \backslash G)_0 = L^2(\Gamma \backslash X)$. On X, the Casimir operator is -2 times the Laplace–Beltrami operator Δ; therefore, 13.4, 13.7, and 13.8
describe the spectral decomposition of $L^2(\Gamma \backslash X)$ with respect to Δ. In view of
13.7(10), this is identical to [36, Thm. 7.3].

(ii) A priori, J_m may depend on m. Using representation theory, we shall see
that the dependence is weak. More precisely, for $z \in (0, 1]$, let $J_m(z)$ be the set
of $j \in \{1, \ldots, l(\delta)\}$ such that $E_j(m, s)$ has a pole at z. Then (see 16.5):

(1) $J_m(z)$ is empty if m is odd, for any $z \in (0, 1]$;

(2) $J_m(z) = J_n(z)$ $(z \in (0, 1), m, n$ even$)$;

(3) $J_m(1) = 0$ for $m \neq 0$.

Note that, by 12.13,

(4) $J_0(1) = \{1, \ldots, l\}.$

In particular, the set J of $z \in (0, 1]$ at which some Eisenstein series $E_j(m, s)$
$(m \in \mathbb{Z}, j = 1, \ldots, l)$ has a pole is finite.

14 Generalities on representations of G

We first review some notions and facts about continuous representations of

$$G = \mathrm{SL}_2(\mathbb{R})$$

in a locally complete topological vector space V. In fact, all we need is the representation by right translations of G in $C^\infty(G)$, $L^2(G)$, $C^\infty(H\backslash G)$, or $L^2(H\backslash G)$ (H closed unimodular subgroup, mainly Γ). Much of this has been encountered earlier, implicitly or explicitly. All this is valid in a much more general framework (see e.g. [7]).

14.1 A continuous representation (π, V) of G into V is a homomorphism of G into the group of automorphisms of V that is continuous; in other words, the map $(g, v) \mapsto \pi(g).v$ is a continuous map of $G \times V$ into V. If V is a Hilbert space then it is said to be *unitary* if $\pi(g)$ ($g \in G$) leaves the scalar product of V invariant. Then the operator norm $\|\pi(g)\|$ is uniformly bounded (by 1), and it is known that continuity already follows from separate continuity:

(1) for every $v \in V$, the map $g \mapsto \pi(g).v$ of G into V is continuous (see e.g. [7, §3] or [10, §VIII.1]).

14.2 The assumption "locally complete" is made to ensure that if $\alpha \in C_c(G)$ then $\int_G \alpha(x)\pi(x).v\,dx$ converges to an element of V. More generally, $\int \mu(x)\pi(x).v$ converges if μ is a compactly supported measure on G (see an important example in 14.4). This allows one to extend π to $C_c^\infty(G)$ by the rule

(1)
$$\pi(\alpha).v = \int_G \alpha(x)\pi(x).v\,dx.$$

If $(\pi, V) = (r, C^\infty(G))$, then $\pi(\alpha)$ is the convolution $*\check{\alpha}$ by $\check{\alpha}$ (see 1.11). This extension satisfies the rule

(2)
$$\pi(\alpha * \beta) = \pi(\alpha) \circ \pi(\beta) \quad \text{if } \alpha, \beta \in C_c^\infty(G)$$

and also if α, β are compactly supported measures.

If $\{\alpha_m\}$ is a Dirac sequence (2.4), then $\pi(\alpha_n).v \to v$, in the topology of V, as $n \to \infty$; see 2.4 for the case of $(r, C^\infty(G))$ and 13.1 for $(r, L^2(\Gamma\backslash G))$.

If (π, V) is unitary, then the adjoint of $\pi(\alpha)$ ($\alpha \in C_c(G)$) is $\pi(\alpha^*)$, as is seen by a simple computation, where $u, v \in V$:

$$(\pi(\alpha)u, v) = \int_G \alpha(x)(\pi(x)u, v)\,dx = \int_G \alpha(x)(u, \pi(x^{-1})v)\,dx$$

$$(\pi(\alpha)u, v) = \int_G (u, \alpha^*(x^{-1})\pi(x^{-1})v)\,dx = (u, \pi(\alpha^*)v).$$

153

14.3 An element $v \in V$ is *differentiable* if the map $g \mapsto \pi(g).v$ of G into V is smooth. The space of differentiable vectors v is denoted V^{∞} and is stable under G. The representation π extends in a natural way to a linear representation of $\mathfrak{g}$ and (more generally) of $\mathcal{U}(\mathfrak{g})$, in End V^{∞}. We shall also denote this representation π, though $d\pi$ would be more correct.

If $(\pi, V) = (r, C^{\infty}(G))$, then every element of V is differentiable and the extension to $\mathcal{U}(\mathfrak{g})$ is the action by left-invariant differential operators considered from Section 2 on. If $(\pi, V) = (r, L^2(G))$ (resp. $(r, L^2(\Gamma \backslash G))$), then a vector is differentiable in this sense if it is represented by a smooth function f such that f and all its derivatives Df $(D \in \mathcal{U}(\mathfrak{g}))$ are square integrable on G (resp. $\Gamma \backslash G$).

If $\alpha \in C_c^{\infty}(G)$ and $D \in \mathcal{U}(\mathfrak{g})$, then $\pi(\alpha).v$ is differentiable since, as can be easily checked, $D(\pi(\alpha).v)) = \pi(D\alpha).v$.

Thus V^{∞} contains $\pi(C_c^{\infty}(G)).V$, which is dense by 14.2 and is called the *Gårding subspace of V*; it is stable under G and $\mathcal{U}(\mathfrak{g})$. (By [16], $V^{\infty} = \pi(C_c^{\infty}(G)).V$, but we shall not need this result.)

14.4 Given $m \in \mathbb{Z}$, we let V_m be the space of vectors of K-type m, that is, of vectors that satisfy

$$(1) \qquad\qquad \pi(k).v = \chi_m(k).v \quad (k \in K);$$

V_m is a closed subspace. There is an idempotent projector π_m of V onto V_m, namely,

$$(2) \qquad\qquad \pi_m(v) = \int_K \chi_{-m}(k).\pi(k).v \, dk.$$

(To check that $\pi_m \circ \pi_m = \pi_m$ and that $\pi_m(v)$ satisfies (1) is immediate and left to the reader.) Let e_m be the measure $\chi_{-m} \, dk$ on K viewed as a measure on G with support on K. Then, by definition, $\pi_m = \pi(e_m)$ in the sense of 14.2. If $\alpha \in I_c^{\infty}(G)$, then $\pi(\alpha)$ commutes with $\pi(k)$ $(k \in K)$, hence $\pi(\alpha)$ leaves the subspace V_m invariant. As a consequence of 14.2 and 14.3,

$$\pi_m(V) \cap V^{\infty} = \pi_m(V^{\infty})$$

is dense in V_m. Let $m \neq n$ $(m, n \in \mathbb{Z})$ and $v \in V_m$. Then

$$\pi_n(v) = \int_K \chi_{-n}(k).\pi(k).v \, dk = \int_K \chi_{-n}(k)\chi_m(k)v \, dk$$

$$= \left(\int_K \chi_{-n}(k)\chi_{-m}(k) \, dk \right).v = 0,$$

which shows that the subspaces V_m $(m \in \mathbb{Z})$ are linearly independent. We let V_K denote their (algebraic) direct sum. Every finite dimensional continuous representation of K splits into 1-dimensional ones, so V_K is also the space of K-finite elements in V – that is, of those v such that $\pi(K).v$ is contained in a finite dimensional subspace of V.

14.5　It is natural to ask whether the "Fourier series" $\sum \pi_m(v)$ converges in some sense to v. In this connection we mention the following fact (for information; it will not be needed). View (π, V) as a representation of K. If v is differentiable with respect to K, then the series $\sum \pi_m(v)$ converges "absolutely" to v. Absolute convergence means that, for each seminorm v in the set of seminorms defining the topology of V, the series $\sum v(\pi_m(v))$ converges (see [7, 3.10]).

If (π, V) is $(r, G(H\backslash G))$ or $(r, L^2(H\backslash G))$, then v_m is just the mth Fourier coefficient of the restriction of v to $x.K$ $(x \subset H\backslash G)$, and the convergence follows from standard facts about Fourier series.

14.6　Admissible representations, $(\mathfrak{g}, K)$-modules.　If V_m is finite dimensional, then $V_m \subset V^\infty$, or even $V_m \subset (C_c^\infty(G).V)$, since $V_m \cap (C_c^\infty(G).V)$ is dense in V_m (14.4). Moreover, $\mathcal{Z}(\mathfrak{g})$ commutes with G on V, hence also with $\pi(e_m)$, and leaves V_m invariant. Therefore V_m consists of K-finite and $\mathcal{Z}$-finite elements.

The representation (π, V) is said to be *admissible* if each V_m is finite dimensional. Then V_K is the space of elements that are K-finite and $\mathcal{Z}$-finite.

We claim that V_K is also the space of elements that are $\mathcal{U}(\mathfrak{k})$-finite, and that V_K is stable under $\mathcal{U}(\mathfrak{g})$. Clearly, any element of V_K is $\mathcal{U}(\mathfrak{k})$-finite. Let now $v \in V$ and assume that $\mathcal{U}(\mathfrak{k}).v = E$ is finite dimensional. Then we can exponentiate the representation of $\mathfrak{k}$ in E; that is, if W generates $\mathfrak{k}$ then, for each $x \in E$, the exponential series

$$\pi(e^{tW}).x = \sum_{n \geqslant 0} \frac{t^n \pi(W^n)}{n!}.x \quad (t \in \mathbb{R})$$

converges to an element of E and hence $E \subset V_K$. Let now $Y \in \mathfrak{g}$ and $v \in V_m$. Then

$$\pi(W).\pi(Y).v = \pi(Y).\pi(W).v + \pi([W, Y]).v \in \mathbb{C}\pi(Y).v + \pi(\mathfrak{g}).v.$$

Using induction on n, one sees similarly that

$$\pi(W^n).\pi(Y).v \subset \mathbb{C}.\pi(Y).v + \pi(\mathfrak{g}).v$$

therefore,

$$\mathcal{U}(\mathfrak{k}).\pi(Y).v \subset \pi(\mathfrak{g}).v.$$

The right-hand side is finite dimensional, so $\pi(Y).v$ is $\mathcal{U}(\mathfrak{k})$-finite and hence belongs to V_K.

A complex vector space E is a $(\mathfrak{g}, K)$-*module* if the following conditions are fulfilled. It is a $\mathfrak{g}$-module and a K-module. It is the union of finite dimensional

K-invariant subspaces on which K acts smoothly. The space E_m is finite dimensional for all $m \in \mathbb{Z}$. For $k \in K$, $X \in \mathfrak{g}$, and $v \in E$, we have $k.X.v = \operatorname{Ad} k(X).k.v$ and finally, on any finite dimensional K-invariant subspace F, the differential of the action of K coincides on F with the restriction to $\mathfrak{k}$ of the given $\mathfrak{g}$-action.

If, moreover, E is finitely generated over $\mathcal{U}(\mathfrak{g})$, then it is called a *Harish-Chandra module*. In particular, if (π, V) is an admissible representation of G, then V_K is a $(\mathfrak{g}, K)$-module.

There is a very close connection between the representations of G in V and those of $\mathcal{U}(\mathfrak{g})$ in V_K when V_K is a Harish-Chandra module, which allows one to algebraize a considerable part of the representation theory. In particular, if $v \in V_K$ then the smallest closed subspace containing $\mathcal{U}(\mathfrak{g}).v$ is G-invariant, whence a bijective correspondence between $\mathcal{U}(\mathfrak{g})$-$K$-invariant subspaces of V_K and closed G-invariant subspaces of V (see [7, 3.17]). Note that, in the case of the right-regular representation on $C^\infty(G)$, this is what was shown in 2.17. In fact, the proof could easily be adapted to cover the more general case.

14.7   The representation (π, V) is (topologically) *irreducible* if V does not contain any nontrivial (i.e. $\neq 0, V$) *closed* subspace invariant under G. If (π, V) is admissible then it is *infinitesimally* irreducible if the $(\mathfrak{g}, K)$-module V_K is (algebraically) irreducible – that is, if V_K does not contain any nontrivial $\mathcal{U}(\mathfrak{g})$-$K$-invariant subspace. The latter is a priori a weaker condition. However, we have just recalled that both conditions are equivalent if V_K is a Harish-Chandra module. In an infinitesimally irreducible representation, C is a multiple of the identity.

14.8   Let (π', V') be another representation of V. A continuous linear map $\sigma: V \to V'$ is an *intertwining operator* if it commutes with G, that is, if

$$\sigma(\pi(g).v) = \pi'(g).\sigma(v) \quad (v \in V, g \in G).$$

In fact, a weaker notion when V_K is a Harish-Chandra module will suffice. We shall say that a linear map $\sigma: \pi(G).V_K \to V'$ is an intertwining operator if it commutes with G. It is not important for us to know whether it extends to V.

14.9 Casimir operator.   The argument given in 13.2 extends to show that, in any unitary representation of G, the operator C is essentially self-adjoint. For a more general reductive group, it proves that if the operator $Q \in \mathcal{Z}(\mathfrak{g})$ is symmetric then it is essentially self-adjoint.

We repeat briefly the argument. Let (π, H) be a unitary representation of G. We take H^∞ as domain of definition of Q. The operator Q on H^∞ commutes with G and hence also with $\pi(\alpha)$, where $\alpha \in C_c^\infty(G)$. By assumption, Q is symmetric:

(1) $$(Qu, v) = (u, Qv) \quad (u, v \in H^\infty).$$

Let T be its adjoint; T is an extension of Q and we must prove that it is self-adjoint. For this it suffices to show that all its eigenvalues on $D(T)$ are real. Let $v \in D(T)$ $(v \neq 0)$ and $\lambda \in \mathbb{C}$ be such that $T.v = \lambda v$. Then

$$T.\pi(\alpha).v = \pi(\alpha)T.v = \lambda\pi(\alpha).v.$$

By using a Dirac sequence, it can be arranged that $\pi(\alpha).v \neq 0$. But it belongs to H^∞ and hence to $D(Q)$, so (1) shows that $\lambda \in \mathbb{R}$.

15 Representations of G

15.1 The principal series. As before (cf. 10.12), $\delta = 0, 1$. Fix a p-pair (P, A). Recall that

$$P = CG.P^\circ, \quad P^\circ = N.A, \quad CG = \mathbb{Z}/2\mathbb{Z}, \quad CG \subset K.$$

Denote by χ_δ the character of P that is trivial on P° and takes the value $(-1)^\delta$ on $-\operatorname{Id}$. The function h_P, restricted to P, is a character that is trivial on $CG.N$. For $s \in \mathbb{C}$, let

$$(1) \qquad\qquad \psi_{P,\delta,s} = \psi_{\delta,s} = \chi_\delta.h_P^{1+s};$$

this is a character of P that is trivial on N. In the C^∞-framework, the principal series representation $(\pi(P, \delta, s), I(P, \delta, s))$ or $(\pi(\delta, s), I(\delta, s))$ is the induced representation $I_P^G(\psi_{\delta,s})$; that is,

$$(2) \quad I(\delta, s) = \{ f \in C^\infty(G) \mid f(p.x) = \psi_{\delta,s}(p).f(x), \ (p \in P, \ x \in G) \},$$

acted upon by right translations.

We shall also view the principal series as a representation in a Hilbert space $H(\delta, s)$, which is the completion of $I(\delta, s)$ with respect to the norm

$$(u, v) = \int_K u(k).\overline{v(k)} \, dk \quad (u, v \in I(\delta, s));$$

that is, $H(\delta, s)$ is the space of functions with square integrable restriction to K that satisfy the condition in (2).

Up to equivalence, this representation is *independent of the p-pair* (P, A). Let (P', A') be another p-pair. There exists $k \in K$ such that $P' = {}^k P$, $A' = {}^k A$ (2.6). Let now $u \in I(P', \delta, s)$. We claim that $l_{k^{-1}}(u) \in I(P, \delta, s)$.

The behavior with respect to CG is, of course, invariant under conjugation. Let $p \in P^\circ$ and $x \in G$. Then, in view of 2.8(5), we have

$$l_{k^{-1}}u(p.x) = u({}^k p.k.x) = h_{P'}({}^k p)^{1+s}.u(k.x) = h_P(p)^{1+s} l_{k^{-1}}u(x),$$

which proves our claim. Moreover, $l_{k^{-1}}$ leaves the scalar product on K invariant, so the norms on $H(P, \delta, s)$ and $H(P', \delta, s)$ are the same. It commutes with right translations and so intertwines $I(P', \delta, s)$ and $I(P, \delta, s)$.

15.2 Proposition. (i) *On $I(\delta, s)^\infty$, the Casimir operator $\mathcal{C} = (s^2 - 1)/2 \operatorname{Id}$.*

(ii) *The space $H(\delta, s)_m$ $(m \in \mathbb{Z})$ is 1-dimensional and spanned by $\varphi_{P,m,s}$ if $m \equiv \delta \bmod 2$ and is zero otherwise.*

(iii) *If $s \notin \mathbb{Z}$, or if $s \in \mathbb{Z}$ and $s \equiv \delta \bmod 2$, then the representation $\pi(\delta, s)$ is irreducible and the $(\mathfrak{g}, K)$-modules $I(\delta, s)_K$ and $I(\delta, -s)_K$ are isomorphic.*

Proof. (i) C commutes with right translations. It suffices therefore to show that $C.h_P^{1+s} = ((s^2 - 1)/2).h_P^{1+s}$. This follows from the argument in the proof of 10.4(v).

(ii) Let $\varepsilon = -\operatorname{Id}$ be the element of CG different from the identity; ε belongs to P and K. Therefore, if $f \in I(\delta, s)_m$ ($f \neq 0$) then $\chi_\delta(\varepsilon) = \chi_m(\varepsilon)$; that is,

(1) $$(-1)^\delta = (-1)^m$$

and hence $m \equiv \delta \bmod 2$. It follows from the definitions that if $m \equiv \delta \bmod 2$ then $\psi_{\delta,s}$ and $\varphi_{P,m,s}$ have the same restriction to P. In view of the relation

$$\varphi_{P,m,s}(p.x) = \varphi_{P,m,s}(p).\varphi_{P,m,s}(x) \quad (p \in P, \ x \in G)$$

(10.2(5)), we see that

$$\varphi_{P,m,s} \in I(\delta, s)_m \quad \text{if } m \equiv \delta \bmod 2.$$

Conversely, let $f \in I(\delta, s)_m$. Then

$$\begin{aligned} f(n.a.k) &= \psi_{\delta,s}(na).\chi_m(k).f(1) \\ &= \varphi_{P,m,s}(n.a.k).f(1) \quad (n \in N, \ a \in A, \ k \in K); \end{aligned}$$

f is therefore a multiple of $\varphi_{P,m,s}$. This proves (ii), at first for $I(\delta, s)$, but since it is dense and admissible in $H(\delta, s)$ we have

(2) $$I(\delta, s)_K = H(\delta, s)_K$$

and of course $I(\delta, s) = H(\delta, s)^\infty$.

(iii) Let us write u_m for $\varphi_{P,m,s}$ ($m \equiv \delta \bmod 2$). We want to prove

(3) $$\mathcal{U}(\mathfrak{g}).u_m = I(\delta, s)_K \quad \text{if } s \notin \mathbb{Z} \text{ or if } s \in \mathbb{Z} \text{ and } s \equiv \delta \bmod 2.$$

Assume it provisorily. This shows first that $I(\delta, s)_K$ is a Harish-Chandra module, which is irreducible as a $(\mathfrak{g}, K)$-module. But this is equivalent to global irreducibility (14.6). There remains to prove (3), which is elementary and may be found in many places. For the convenience of the reader, we give a proof, following [57, 5.6.1].

Let H, E, F be as in 2.1. As in 2.18, let $g \in SL_2(\mathbb{C})$ be such that

$$\operatorname{Ad} g(H) = g.H.g^{-1} = i.W,$$

and let $Y = \operatorname{Ad} g(E)$ and $Z = \operatorname{Ad} g(F)$. Then

(4) $$[iW, Y] = 2Y, \quad [iW, Z] = -2Z, \quad [Y, Z] = iW,$$

(5) $$C = -\tfrac{1}{2}W^2 + iW + 2ZY = -\tfrac{1}{2}W^2 - iW + 2YZ,$$

as pointed out in 2.18(1) and (3). The relations (4) imply, as in 2.18(5), that

(6) $\qquad Y(I(\delta, s)_m) \subset I(\delta, s)_{m+2}, \qquad Z(I(\delta, s)_m) \subset I(\delta, s)_{m-2}.$

Hence it suffices, in order to establish (3), to show that Y and Z are injective on $I(\delta, s)_m$ if $s \notin \mathbb{Z}$ or if $s \in \mathbb{Z}$ and $s \equiv \delta \bmod 2$. (By (ii), we may assume that $m \equiv \delta \bmod 2$.) From (i) and (5) we obtain

(7)
$$2ZY.u_m = \left(\frac{s^2 - 1}{2} - \frac{m^2}{2} - m\right).u_m = \frac{1}{2}(s^2 - (m + 1)^2).u_m,$$
$$2YZ.u_m = \left(\frac{s^2 - 1}{2} - \frac{m^2}{2} + m\right) = \frac{1}{2}(s^2 - (m - 1)^2).u_m.$$

The right-hand sides do not equal zero if $s \notin \mathbb{Z}$ or if $s \in \mathbb{Z}$ and $s \equiv \delta \bmod 2$ (since $m \equiv \delta \bmod 2$), and the first assertion of (iii) follows. To prove the second one, we shall follow [56, pp. 49–51].

Fix $m \equiv \delta \bmod 2$. We define inductively a new basis v_n ($n \in \mathbb{Z}$, $n \equiv \delta \bmod 2$) of $I(\delta, s)$ by the rule

(8) $\qquad\qquad\qquad\qquad\qquad v_m = u_m,$

(9)
$$v_{m+2j+2} = 2(s + m + 2j + 1)^{-1}.Y.v_{m+2j},$$
$$v_{m-2j-2} = 2(s - m + 2j + 1)^{-1}.Z.v_{m-2j}.$$

for $j \in \mathbb{N}$. This is a basis in view of (ii) and the fact that Y and Z are injective on $I(\delta, s)_n$ ($n \in \mathbb{Z}$, $n \equiv \delta \bmod 2$), as was proved to establish the irreducibility of $I(\delta, s)$. We claim:

(10) $\qquad Y.v_{m+2j} = \frac{1}{2}(s + m + 2j + 1).v_{m+2j+2} \quad (j \in \mathbb{Z});$

(11) $\qquad Z.v_{m+2j} = \frac{1}{2}(s - m - 2j + 1).v_{m+2j-2} \quad (j \in \mathbb{Z}).$

For $j \geqslant 0$, (10) follows by definition of the v_{m+2j}. For $j > 0$, by construction we have
$$Y.v_{m-2j} = 2(s - m + 2j + 1)^{-1}Y.Z.v_{m-2j+2},$$
and then (10) follows from (7) for $j < 0$. The proof of (11) is similar. Recall (2.12(1)) that

(12) $\qquad\qquad\qquad\qquad iW.v_n = -m.v_n \quad (n \in \mathbb{Z}).$

It is clear that, if s is replaced by $-s$ in (9), then (10), (11), and (12) are true with s replaced by $-s$. This would be our construction in $I(\delta, -s)_K$, whence the second part of (iii).

15.3 Corollary. *Fix $\delta \in \{0, 1\}$ and $m, n \in \mathbb{Z}$. There exists a function $q_{\delta,m,n}(g, s)$ on $G \times \mathbb{C}$ which, for fixed $g \in G$, is entire in s, and such that*

(1) $\qquad\qquad\qquad\qquad (r_g u_m)_n = q_{\delta,m,n}(g, s).u_n.$

If $m \equiv n \equiv \delta \bmod 2$, *and if* $s \notin \mathbb{Z}$ *or if* $s \in \mathbb{Z}$ *and* $s \equiv \delta \bmod 2$, *then* $q_{\delta,m,n}(g,s)$ *is not identically zero on* G.

Proof. By definition, the value of the left-hand side at $x \in G$ is

$$(2) \qquad I(x) = \int_K \chi_{-n}(k') . \varphi_{P,m,s}(x.k'.g) \, dk'$$

$$= \int_K \chi_{-n}(k') . h_P^{1+s}(x.k'.g) \varphi_{P,m}(x.k'.g) \, dk.$$

Write

$$(3) \qquad x = q.a.l \quad (q \in N, \ a \in A, \ l \in K),$$

and change to the variable $k = l.k'$. We have

$$h_P(x.k', g) = h_P(a.k.g) = h_P(a).h_P(k.g) = h_P(x).h_P(k.g),$$
$$\chi_{-n}(k') = \chi_n(l).\chi_{-n}(k);$$

therefore,

$$(4) \qquad I(x) = \varphi_{P,n,s}(x) \int_K \chi_{-n}(k) h_P^{1+s}(k.g) \varphi_{P,m}(k.g) \, dk,$$

which yields (1) if we set

$$(5) \qquad q_{\delta,m,n}(g,s) = \int_K \chi_{-n}(k) . h_P^{1+s}(k.g) . \varphi_{P,m}(k.g) \, dk.$$

This function is clearly entire in s. Assume now that $s \notin \mathbb{Z}$ or that $s \in \mathbb{Z}$ and $s \equiv \delta \bmod 2$. Then $I(\delta, s)$ is irreducible, so the set of translates $r_g(u_m)$ spans a dense subspace of $I(\delta, s)$ and its projection on $I(\delta, s)_n$ is not zero; that is, the projection of $r_g.u_m$ on $I(\delta, s)_n$ is a nonzero multiple of u_n for a suitable g. This proves the last assertion.

15.4 Lemma. *Let* (P, A) *be a p-pair and let* $g \in G$. *Given* $k \in K$, *let* p_k *and* l_k *be the unique elements in* P° *and* K *respectively such that*

$$(1) \qquad k.g = p_k.l_k.$$

Then

$$(2) \qquad \int_K u(k) \, dk = \int u(l_k) . h_P^2(p_k) \, dk.$$

Proof. Let $v \in C_c(P)$ be such that

$$(3) \qquad \int_P v(p) \, d_l p = 1,$$

and let $v.u$ be the function on G defined by $v.u(p.k) = v(p).u(k)$ ($p \in P^\circ$, $k \in K$). Recalling that $dx = d_l p \, dk$ is a Haar measure on G and that G is unimodular, we have

$$\int_K u(k) \, dk = \int_G vu(x) \, dx = \int_G vu(xg) \, dx$$

$$= \int_G vu(pkg) \, d_l p \, dk = \int_G vu(pp_k l_k) \, d_l p \, dk$$

$$= \int_K u(l_k) \, dk \int_P v(pp_k) \, d_l p.$$

However,

$$\int_P v(pp_k) \, d_l p = \int v(p) h_P^2(p_k) \, d_l p$$

(see 2.9) and so, by changing the order of integration, we obtain

$$\int_K u(k) \, dk = \int v(p) \, d_l p \int_K u(l_k) h_P^2(p_k) \, dk;$$

the lemma now follows from (3).

15.5 Proposition.   *The principal series $H(\delta, s)$ is contragredient to $H(\delta, -s)$. If $s \in i\mathbb{R}$ then $H(\delta, s)$ is unitary and equivalent to $H(\delta, -s)$, and the functions $\varphi_{P,m,s}$ ($m \equiv \delta \bmod 2$) form an orthonormal basis.*

To prove the first assertion, it suffices to exhibit a nondegenerate continuous bilinear form $\langle \, , \, \rangle$ on $H(\delta, s) \times H(\delta, -s)$ that is G-invariant. Let

$$(1) \qquad \langle u, v \rangle = \int_K u(k).v(k) \, dk \qquad (u \in H(\delta, s), \ v \in H(\delta, -s)).$$

We first show that this form is G-invariant. It suffices to prove it for $u \in I(\delta, s)$ and $v \in I(\delta, -s)$. Let $g \in G$. We use the notation of 15.4(1) and write

$$\langle r_g u, r_g v \rangle = \int_K u(k.g).v(k.g) \, dk = \int_K u(p_k.l_k).v(p_k.l_k) \, dk$$

$$= \int_K u(l_k)v(l_k)h_P^2(p_k) \, dk = \int u(k).v(k) \, dk = \langle u, v \rangle$$

by 15.4. Let now $V \subset H(\delta, s)$ be the subspace orthogonal to $H(\delta, -s)$ with respect to $\langle \, , \, \rangle$. The subspace V is G-invariant and closed, so if it is not zero then $V_m \neq 0$ for some m; that is, V contains a function $\varphi_{P,m,s}$ for some $m \equiv \delta \bmod 2$ (15.2). But $H(\delta, -s)$ contains $\varphi_{P,-m,s}$ and

$$\langle \varphi_{P,m,s}, \varphi_{P,-m,-s} \rangle = \int_K \chi_m(k).\chi_{-m}(k) \, dk = 1,$$

a contradiction, so $V = 0$. This also proves that no nonzero element of $H(\delta, -s)$ is orthogonal to $H(\delta, s)$, so $\langle\ ,\ \rangle$ is nondegenerate.

Let now $s \in i\mathbb{R}$ and $u, v \in H(\delta, s)$. Then the previous computation shows that the scalar product

$$(2) \qquad\qquad (u, v) = \int_K u(k).\overline{v(k)}\, dk$$

is G-invariant and hence the representation is unitary. That $H(\delta, s)$ is equivalent to $H(\delta, -s)$ is clear if $s = 0$. If $s \in i\mathbb{R}$, ($s \neq 0$) then it follows from 15.2.

The scalar product is given by (2), so

$$(\varphi_{P,m,s}, \varphi_{P,n,s}) = \int_K \varphi_{P,m,s}(k)\bar{\varphi}_{P,n,s}(k)\, dk = \int_K \chi_m(k).\chi_{-n}(k)\, dk = \delta_{m,n},$$

which proves the last assertion.

Remark. The proofs of 15.4 and 15.5 are in substance those of Theorems 2 and 3 in [40, §III.2].

15.6 By 15.2, the only candidates for reducibility are the $H(\delta, s)$ for $s \in \mathbb{Z}$ ($s \not\equiv \delta \bmod 2$); they are, in fact, reducible. We only state the results, which are not really needed here (except for one case discussed in 15.7), referring to the literature for the proofs (see [57, 5.6; 56, Chap. 1; 37, §II.5, §VI.1]).

In each dimension, $\mathrm{SL}_2(\mathbb{C})$ has – up to equivalence – one irreducible representation acting on the space of homogeneous polynomials on $\mathbb{C}^2$ of some degree n, to be denoted F_n. It has degree $n + 1$ and the Casimir operator has eigenvalue $(n^2/2) + n$; it also has the K-types m, where $m \equiv n \bmod 2$ and $m \in [-n, n]$, and is self-contragredient.

For $n \in \mathbb{Z} - \{0\}$, G has an irreducible representation D_n, a "discrete series representation" that is square integrable – that is, it can be realized on a closed G-invariant subspace of $L^2(G)$. For $|n| \geq 2$, this representation is even integrable, that is, it belongs to $L^1(G)$. A definition will be recalled in 15.10. The eigenvalue of $\mathcal{C}$ on D_n is $(n^2 - 1)/2$. The K-types of D_n are the $m \in \mathbb{Z}$, where $n \equiv m \bmod 2$, $\mathrm{sgn}\, n = \mathrm{sgn}\, m$, and $|m| > |n|$.

Consider now $H(\delta, s)$ for $s \in \mathbb{Z}$ ($s \not\equiv \delta \bmod 2$, $s \neq 0$). Let n be a strictly positive integer. Then $H(\delta, n)$ contains $D_n \oplus D_{-n}$ and the quotient is F_{n-1}, whereas $H(\delta, -n)$ contains F_{n-1} and the quotient by the latter is $D_n \oplus D_{-n}$.

If $s = 0$, then $I(1, 0)$ is unitary and decomposes into the sum of two representations, denoted $D_{+,0}$ and $D_{-,0}$, called *limits of discrete series*.

15.7 We now consider $I(0, 1)$. The K-types are even. We see from 15.2(6) and (7) that the subspace D_1 (resp. D_{-1}) spanned by the u_m (resp. u_{-m}) for $m \geq 2$ are irreducible $(\mathfrak{g}, K)$-modules. We claim that the $r_g(u_0)$ $(g \in G)$ span a dense subspace of $I(0, 1)$. Recall (15.1) that $u_0 = h_P^2$. The latter is not invariant under G, hence there exist $g \in G$ and $m \in 2\mathbb{Z} - \{0\}$ such that the projection of $r_g(u_0)$ on $I(0, 1)_m$ is not zero; that is,

$$\int \chi_{-m}(k) h_P^2(g.k) \, dk \neq 0.$$

By going over to complex conjugates, we see that the projection of $r_g(u_0)$ on $I(0, 1)_{-m}$ is also not equal to zero. We then use the previous remarks on D_1 and D_{-1} to conclude that, given m even, there exists g such that the projection of $r_g(u_0)$ on $I(0, 1)_m$ is not zero. As a consequence, we can complete 15.3 in this case by part (i) of the following lemma.

Lemma. (i) *The function* $q_{0,m,n}(g, 1)$ *is not identically zero on G if either $m = 0$ and n is even, or if m, n are even, of the same sign, and not zero.*

(ii) *The only* $(\mathfrak{g}, K)$-*module quotient of* $I(0, 1)_K$ *that contains the trivial representation* F_0 *is the quotient* $I(0, 1)_K / (D_1 \oplus D_{-1})$, *which is equal to* F_0.

Proof of (ii). Let V be a $(\mathfrak{g}, K)$-module quotient of $I(0, 1)$, and let W be the kernel of the projection σ. The only possibilities for W are 0, D_1, D_{-1}, and $D_1 \oplus D_{-1}$. In the first three cases, the only K-invariant vectors are the multiples of $\sigma(u_0)$, but the translates of $\sigma(u_0)$ generate V by the previous discussion. In the fourth case, the quotient is indeed 1-dimensional and isomorphic to F_0.

15.8 The irreducible unitary representations of G are, up to equivalence:

(a) the discrete series D_n ($n \in \mathbb{Z} - \{0\}$);
(b) the unitary principal series $H(\delta, s)$, where $s \in i\mathbb{R}$ and $(\delta, s) \neq (0, 0)$;
(c) the two constituents $D_{+,0}$ and $D_{-,0}$ of $H(0, 0)$, the *limits of discrete series*;
(d) the representations $I(0, s)$ for s real ($s \in (0, 1)$), called the *complementary series*;
(e) the trivial 1-dimensional representation.

If to this we add:

(f) the finite dimensional irreducible representations F_n ($n \geqslant 1$) and
(g) the irreducible principal series $I(\delta, s)$ – that is, $s \notin \mathbb{Z}$, or $s \in \mathbb{Z}$ and $s \equiv \delta \bmod 2$, $\mathcal{R}s > 0$,

then the underlying $(\mathfrak{g}, K)$-modules of K-finite vectors are, up to equivalence, all irreducible admissible $(\mathfrak{g}, K)$-modules. This is the Langlands classification in this case (cf. [57]). To go from there to a classification of topologically irreducible representations in topological vector spaces, one must investigate the various possible completions – that is, the smallest closed G-invariant subspace – in some topological vector G-space whose smooth K-finite vectors realize the given $(\mathfrak{g}, K)$-module; there may be several, up to equivalence. However, this ambiguity does not exist for a realization in a unitary G-module (see [57, XI] for a discussion of this matter). The set of unitary equivalence classes of irreducible unitary representations of G is denoted $\hat{G}$.

15.9 Direct integral of unitary principal series. This is the only example of direct integral of representations that we shall need, so we limit ourselves to this case. For a general definition, see for example [57, 14.8].

We fix a cuspidal p-pair (P, A), δ and write $H(P, \delta, s)$ for the Hilbert space model of the principal series $I(P, \delta, s)$. We always assume s to be purely imaginary, so that $H(P, \delta, s)$ is a unitary representation (15.5).

In this section we write $u_m(s)$ for $\varphi_{P,m,s}$. Recall that the $u_m(s)$ ($m \equiv \delta \bmod 2$) form an orthonormal basis of $H(P, \delta, s)$ (15.5). We let $(\ ,\ )_s$ and $\| \cdot \|_s$ denote the scalar product and norm on $H(P, \delta, s)$, respectively.

We now consider the family of Hilbert spaces $H(P, \delta, s)$ ($s \in i\mathbb{R}^+$). Intuitively, this family should be viewed as a space endowed with a projection p on $i\mathbb{R}^+$ and fiber $p^{-1}(s)$ over s equal to $H(P, \delta, s)$. By definition, a *section* $c: s \mapsto c(s)$ of the family assigns to $s \in i\mathbb{R}^+$ an element $c(s) = \sum_m c_m(s).u_m(s) \in H(\delta, s)$ such that, for each m, the function $s \mapsto c_m(s)$ is measurable. The section is square integrable if $\int_{i\mathbb{R}^+} \|c(s)\|_s^2 \, d\tau < \infty$. The space of square integrable sections, modulo the equivalence relation $c = e$ if $\int_{i\mathbb{R}^+} \|c(s) - e(s)\|_s^2 \, d\tau = 0$ and endowed with the scalar product $\langle c, d \rangle = \int_{i\mathbb{R}^+} (c(s), d(s))_s \, d\tau$, is a Hilbert space that we denote $\int_{i\mathbb{R}^+} H(\delta, s) \, d\tau$, the *direct integral* $\mathbb{H}_\delta$ of the $H(\delta, s)$. We let G act on the space of sections via the given representation on each fiber, that is, $(g.c)(s) = g.c(s)$ ($s \in i\mathbb{R}^+$). The group G leaves the space of square integrable sections invariant. The representation in $\int_{i\mathbb{R}^+} H(P, \delta, s) \, d\tau$ so defined is unitary, and called the *direct integral of the unitary principal series*.

These direct integrals are independent of P, up to natural isomorphisms, but we will have to keep track of cusps later. We therefore let

$$\mathbb{H}_{\delta,j} = \int_{i\mathbb{R}^+} H(P_j, \delta, s) \, d\tau \quad (\delta = 0, 1, \ j \in I(\delta)).$$

It is naturally isomorphic to $\mathbb{H}_\delta$.

15.10 To end this section, we give a natural realization of D_m in $(r, L^2(G))$. In fact, 4.5 already contains implicitly such a realization in $(l, L^2(G))$. The present set-up, however, favors the right-regular representation, so we first modify slightly the underlying construction and combine it with the inversion on G so as to switch sides.

Given a function f on D, we now let $\tilde{f}$ be defined by

(1) $$\tilde{f}(x) = f(x^{-1}.0).\mu(x^{-1}, 0)^{-m};$$

we let x operate on the functions on D by

(2) $$(x \circ f)(w) = f(x^{-1}.0).\mu(x^{-1}, 0)^{-m}.$$

Then $\tilde{f}$ is of left K-type m and

(3) $$(x \circ f) = r_x f.$$

Moreover, if $f = w^n$ (resp. $\bar{w}^n$), then $\tilde{f}$ is of right K-type $m+2n$ (resp. $-m-2n$). This follows by computations similar to those used in 5.13–5.15, or can be deduced from those and the fact that the inversion changes l_g to r_g; both functions are still eigenfunctions of $\mathcal{C}$ with eigenvalue $(m^2/2) - m$. For $m \geqslant 0$, we let E_m be the vector subspace of $C^\infty(G)$ spanned by the $\tilde{f}$, where f is a polynomial in w and E_{-m} is the space spanned by the complex conjugate functions. The spaces E_m and E_{-m} consist of functions that are bounded for $m \geqslant 0$, square integrable for $m \geqslant 2$, and integrable for $m \geqslant 4$ (4.5).

Then, for $m \geqslant 1$, E_{m+1} (resp. E_{-m-1}) is a realization of $D_{m,K}$ (resp. $D_{-m,K}$). The closure of the smallest right G-invariant subspace containing it is a realization of D_m (resp. D_{-m}) as a unitary representation – square integrable if $m \geqslant 1$, integrable if $m \geqslant 3$.

16 Spectral decomposition of $L^2(\Gamma\backslash G)$:
the discrete spectrum

Here again, we let $H = L^2(\Gamma\backslash G)$, $^\circ H = {}^\circ L^2(\Gamma\backslash G)$, and V be the orthogonal complement of $^\circ H$ in H.

The space H is the Hilbert direct sum of the subspace H_m ($m \in \mathbb{Z}$), so we already have a spectral decomposition of H with respect to C. The list in 15.8 shows that there are at most two inequivalent irreducible unitary representations of G with the same Casimir operator, so there is in principle not much difference between the decompositions with respect to C and to G. But we want to express the latter in terms of representations.

16.1 Lemma. *Let (π, E) be a unitary representation of G. Assume the existence of a Dirac sequence $\{\alpha_n\}$ ($n \in \mathbb{N}$) with compact operators $\pi(\alpha_n)$. Then E is a Hilbert sum of irreducible representations, with finite multiplicities.*

Proof. This is rather elementary and well-known. For the convenience of the reader we reproduce one proof, following 5.8 and 5.9 in [7].

(a) We show first that a closed G-invariant subspace $W \neq 0$ contains a closed G-invariant subspace that is minimal among nonzero closed G-invariant subspaces. By 13.1 and 14.2, there exists a j such that $\pi(\alpha_j)|_W \neq 0$. Let c be a nonzero eigenvalue of $\pi(\alpha_j)$ in W, and let $M \neq 0$ be the corresponding (finite dimensional) eigenspace in W. Let N be a minimal element among the nonzero intersections of M with the closed G-invariant subspaces of W. Let $v \in N - \{0\}$ and P be the smallest closed G-invariant subspace containing v. By construction, $P \cap M = N$. Let us show that P is minimal. Let Q be a closed G-invariant subspace of P, and let R be its orthogonal complement in P. Then $P = Q \oplus R$. The spaces P, Q, R, and M are invariant under $\pi(\alpha_j)$, hence $N = P \cap M = Q \cap M \oplus R \cap M$. Therefore either $N = Q \cap M$ with $P = Q$ and $R = \{0\}$, or $N = R \cap M$ with $P = R$ and $Q = \{0\}$.

(b) We next prove that E has a discrete decomposition in closed irreducible G-invariant subspaces. Let S be the set consisting of the nonzero closed G-invariant subspaces admitting, and endowed with, a discrete decomposition. By (a), S is not empty. This set is partially ordered by a relation $X \leqslant Y$ if the space of X is contained in that of Y and the discrete decomposition of Y extends that of X; this is an inductive ordering. Let W be the space underlying a maximal element. If $W \neq E$ then we could add to W a minimal closed nonzero G-invariant subspace in the orthogonal complement of W, which exists by (a), whence a contradiction. Thus $W = E$.

(c) There remains to show that the multiplicities are finite. Let (σ, F) be an irreducible unitary representation occurring in E, and let n_σ be its multiplicity. Let j be such that $\sigma(\alpha_j)$ is not zero on F. Let again c be a nonzero eigenvalue of $\sigma(\alpha_j)$ and m its multiplicity. Then the dimension of the eigenspace of $\pi(\alpha_j)$ in E with eigenvalue c is finite on the one hand, and at least equal to $n_\sigma.m$ on the other. Hence n_σ is finite.

16.2 Theorem. *The space $^\circ H$ decomposes into a Hilbert direct sum of closed irreducible G-invariant subspaces with finite multiplicities.*

Proof. In the right regular representation $(r, {}^\circ H)$, the endomorphism $\pi(\alpha)$ is $*\check{\alpha}$ ($\alpha \in C_c^\infty(G)$). It is compact by 9.3, so we can apply the previous lemma.

16.3 We saw earlier (8.9) that the Poincaré series associated to L^1-functions that are K-finite on both sides and $\mathcal{Z}$-finite are cusp forms and therefore belong to $^\circ H$. In fact, more precisely, the formation of Poincaré series provides an intertwining operator from the discrete series D_m ($|m| \geqslant 3$) to $^\circ H$, or rather an intertwining operator from the space of G-translates of elements in $D_{m,K}$ to $^\circ H$. To see this, we start from the realization of $D_{m,K}$ in $L^1(G)$ given in 15.10. For $m \geqslant 3$, it has a basis $\{f_{m,n}\}$ ($n \geqslant 0$), where

$$f_{m,n}(x) = (x^{-1}.0)^n . \mu(x^{-1}, 0)^{-m-1} \quad (n \in \mathbb{Z})$$

has left K-type $m+1$ and right K-type $m+1+2n$. The construction of Poincaré series supplies a linear map $p_m: F_{m,K} \to {}^\circ H$ that sends $f_{m,n}$ to the Poincaré series $p_{f_{m,n}}$. This map commutes with right translations and may therefore be viewed as an intertwining operator from $G.D_{m,K}$ to $^\circ H$.

Similarly, $D_{-m,K}$ has a basis consisting of the $\bar{f}_{m,n}$, and we have an intertwining operator p_{-m} mapping $\bar{f}_{m,n}$ to the Poincaré series $p_{\bar{f}_{m,n}}$.

There remains to investigate V. We first consider the residual spectrum.

16.4 Eisenstein transform. We fix a cuspidal p-pair (P, A) and again write $I(\delta, s)$ for $I(P, \delta, s)$.

The space $I(\delta, s)_K$ has a basis consisting of the $\varphi_{P,m,s}$ ($m \equiv \delta \bmod 2$) by 15.2(ii). For $\mathcal{R}s > 1$, we let

$$\mathbb{E}(P, s): I(\delta, s)_K \to C^\infty(\Gamma \backslash G)$$

be the linear map that assigns $E(P, m, s)$ to $\varphi_{P,m,s}$. This map commutes with right translations and hence extends to an intertwining operator from $G.I(\delta, s)_K$ to $C^\infty(\Gamma \backslash G)$. By analytic continuation, it extends meromorphically to $\mathbb{C}$; it will be called the *Eisenstein transform*.

Assume that $I(\delta, s)$ is irreducible (cf. 15.2(iii)). Fix $z \in \mathbb{C}$. From 15.3, we see that if $E(P, m, s)$ is holomorphic at z for some $m \equiv \delta \bmod 2$, then $E(P, n, s)$ is holomorphic at z for all $n \equiv m \bmod 2$. In the case of $I(0, 1)$, Lemma (i) in 15.7 shows similarly that if $E(P, 0, s)$ is holomorphic at z then so is $E(P, m, s)$ for all even m.

16.5   The residual spectrum.   Let $z \in (0, 1)$ and assume that $E(P, s, m)$ has a simple pole at z ($m \equiv \delta \bmod 2$). Then so do all $E(P, s, n)$ ($n \equiv m \bmod 2$). We define the residue $\mathbb{E}'(P, z)$ of $\mathbb{E}(P, s)$ at z as the linear map $I(\delta, z)_K \to C^\infty(\Gamma \backslash G)$ that sends $\varphi_{P,m,s}$ to

$$E'(P, m, z) = \lim_{s \to z}(s - z)E(P, m, s).$$

But the $E'(P, m, z)$ are square integrable (11.11), so $\mathbb{E}'(P, z)(G.I(\delta, z)_K)$ is unitarizable; it is a copy of $I(\delta, z)$, so the latter is unitarizable, too. By 15.8, it belongs to the complementary series if $\delta = 0$, and this is not possible for $\delta = 1$. As a consequence, $E(P, m, s)$ is holomorphic on $(0, 1)$ if m is odd. This also holds, for the same reason, at $z = 1$. Let again $z \in (0, 1) \cap J_m$. Then the $E'_j(m, z)$ span a copy, say $M_{j,K}$, of the $(\mathfrak{g}, K)$-module $H(0, z)_K$ if the $E_j(m, s)$ indeed have a pole there. Then the smallest closed subspace of V containing that pole is a copy $M_{j,z}$ of $H(0, z)$. The $M_{j,z}$ may not be linearly independent. The linear dependence relations between them are those of the $E'_k(m, z)$ for any even m. In fact, if there is a linear relation between those functions then it is also true for their translates by any $g \in G$, and hence for the components of type n of those translates, and finally for the $E'_j(n, z)$ by 15.3. Let then d_z be the dimension of the space spanned by the $E'_k(m, z)$ for a given m. We just saw that this dimension is independent of m. This also proves 13.9(1) and (2).

There remains to consider the residues at 1. By the foregoing, only $I(0, 1)$ can contribute. By 12.13, each $E_j(0, s)$ has a pole at 1, with residue the constant function equal to $\mathrm{vol}(\Gamma \backslash G)^{-1}$. Now 15.7 implies that the kernel of the intertwining operator $\mathbb{E}'(P_j, 1)$ is $D_1 \oplus D_{-1}$, so the Eisenstein series $E_j(m, s)$ are holomorphic at 1 for m even ($m \neq 0$), which concludes the proof of 13.9(3).

In the sequel we let J be the set of $z \in (0, 1]$ at which some Eisenstein series $E_j(m, s)$ has a pole ($m \in \mathbb{Z}$, $j = 1, \ldots, l$); J is finite and contains 1. Let also $J' = J \cap (0, 1)$.

For $z \in J'$, we let M_z be the span of the $M_{m,j}$. We claim that its closure $\mathrm{cl}(M_z)$ is the eigenspace $E(\mathcal{C}, (z^2 - 1)/2)$ of $\mathcal{C}$ in V for the eigenvalue $(z^2 - 1)/2$, to which the space M_z clearly belongs. On the other hand, $E(\mathcal{C}, (z^2 - 1)/2)$ is closed and is the Hilbert direct sum of its components of K-type m ($m \in \mathbb{Z}$); 13.8 shows that each such component is equal to $M_{z,m}$ and hence belongs to M_z. The space M_z is spanned by d_z linearly independent copies of $H(0, z)$. They may not be orthogonal, but nevertheless $\mathrm{cl}(M_z)$ is the orthogonal direct sum of d_z copies of $H(0, z)$. To see this, use induction on d_z and apply the induction assumption to the quotient of $\mathrm{cl}(M_z)$ by one of the $M_{j,z}$.

For $z = 1$, we let M_z be the space of constant functions. For $z \neq z'$ ($z, z' \in J$), the spaces M_z and M'_z are orthogonal. Hence the direct sum of the spaces $\mathrm{cl}(M_z)$ ($z \in J$) is closed; we denote it by H_{rs}. We have proved the following.

16.6 Theorem. *Let J be the set of $z \in (0, 1]$ at which some Eisenstein series $E_j(m, s)$ has a pole. Then J is finite and contains 1. The space H_{rs} is the direct sum of the constants and of finitely many copies of the complementary unitary principal series $H(0, z)$ ($z \in J$, $z \neq 1$); it is the full discrete spectrum of C in V.*

17 Spectral decomposition of $L^2(\Gamma\backslash G)$: the continuous spectrum

There remains to investigate the orthogonal complement of $^\circ H \oplus H_{rs}$ in H, to be denoted H_{ct} (ct for continuous). It is the orthogonal direct sum of the spaces $H_{ct,m}$ considered in Section 13, and the main results there already provide a spectral decomposition. But we want to express the complement in terms of representations, by establishing an isomorphism of unitary G-modules between H_{ct} and an orthogonal sum of direct integrals of unitary principal series (15.9). There are two natural transformations to relate the two, which are inverse to one another: the *Eisenstein transform*, used in [36], and the "Laplace transform" introduced in [23, 24], to be called here the *Godement transform*. The former suffices to establish the isomorphism, and we shall treat it first.

17.1 Some notation. We let H_{ct} be the orthogonal complement of H_{rs} in V, that is, of $^\circ H \oplus H_{rs}$ in H. As usual (12.0), we write H_{ct} as an orthogonal sum

$$(1) \qquad\qquad H_{ct} = {}_0H_{ct} + {}_1H_{ct}.$$

We also have

$$(2) \qquad\qquad {}_\delta H_{ct,m} = H_{ct,m} \quad \text{if } m \equiv \delta \bmod 2,$$

$$(3) \qquad\qquad {}_\delta H_{ct} = \bigoplus_{m\equiv\delta(2)} H_{ct,m}.$$

As in 16.6, J is the (finite) set of points in $(0,1]$ at which some Eisenstein series $E_i(m,s)$ has a pole. Finally, set

$$(4) \qquad {}_\delta Q_j = \{ u \in {}_\delta C_c^\infty(\Gamma_{P_j} N_j \backslash G) \mid \hat{u} \text{ is zero on } J \}, \qquad Q_j = {}_0Q_j + {}_1Q_j.$$

17.2 Proposition. *Let Q be the vector subspace of $C_c^\infty(G)$ spanned by the union of the Q_j, and let E_Q be the vector subspace of V spanned by the E-series E_u ($u \in Q$).*

(i) *The E-series E_u ($u \in C_c^\infty(\Gamma_{P_j}.N_j\backslash G)$) belongs to H_{ct} if and only if $u \in Q_j$.*

(ii) *E_Q is dense in H_{ct}. The restrictions to $i\mathbb{R}$ of the function $\hat{u}(s)$ ($u \in Q_j$) are dense in $L^2(i\mathbb{R})$.*

Proof. (i) follows from 12.6(3).

From 17.1(2) and (3) we see that, in order to prove the first assertion of (ii), it suffices to show that $E_Q \cap H_{ct,m}$ is dense in $H_{ct,m}$.

171

The space E_Q is dense in V (13.5); therefore, $E_{Q,m}$ is dense in V_m. We have $V_m = H_{ct,m} \oplus H_{rs,m}$. The space $H_{rs,m}$ is finite dimensional and is spanned by the residues $E_i'(m, z)$ of the Eisenstein series $E_i(m, s)$ that have poles on $(0, 1]$ (13.8). Each such residue defines a linear form $E_u \mapsto (E_u, E_j'(m, z))$ on $E_{Q,m}$, so $E_{Q,m} \cap H_{ct,m}$ has codimension in $E_{Q,m}$ at most equal to dim $H_{rs,m}$, which is equal to the codimension of $H_{ct,m}$ in V_m. Since the sum of a finite dimensional space and a closed subspace is closed, our assertion follows.

The second part of (ii) is a consequence of the following lemma.

17.3 Lemma. *Let $D_\sigma = \sigma + i\mathbb{R} \subset \mathbb{C}$ ($\sigma \in \mathbb{R}$), and let E be a finite set of points in $\mathbb{C}$ outside D_σ. Then the restrictions to D_σ of the PW functions on $\mathbb{C}$ that vanish on E are dense in $L^2(D_\sigma)$.*

Proof. We note first that the PW functions form an algebra, and also a module over the polynomials on $\mathbb{C}$. This can be seen from the characterization of PW functions (see 12.4) or from the definition, since the Fourier–Laplace transform maps convolution to product and differentiation by $\frac{d}{dt}$ ($t = h_P$) to multiplication by s.

Let P be a polynomial on $\mathbb{C}$ vanishing exactly on E, and let $\psi = e^{s^2} . P$. It suffices to show that the set of restrictions to D_σ of the functions belonging to the ideal of PW functions generated by ψ is dense in $L^2(D_\sigma)$. If $f \in L^2(D_\sigma)$ is orthogonal to all such functions, then $f\bar{\psi}$ is orthogonal to the restrictions of all PW functions and hence is zero; therefore, f is zero.

17.4 The Eisenstein transform. Let $u \in {}_\delta Q_j$ and $v \in {}_\delta Q_k$ be right K-finite. It follows from 13.8(1), (5), and (9) that we have

$$(1) \qquad E_u = (2\pi)^{-1} \sum_{m,n} \int_0^\infty (E_u, E_n(m, i\tau)) E_n(m, i\tau) \, d\tau,$$

$$(2) \qquad (E_u, E_v) = (2\pi)^{-1} \sum_{m,n} \int_0^\infty (E_u, E_n(m, i\tau))(E_n(m, i\tau), E_v) \, d\tau,$$

$$(3) \qquad \|E_u\|^2 = (2\pi)^{-1} \sum_{m,n} \int_0^\infty |(E_u, E_n(m, i\tau))|^2 \, d\tau.$$

In these sums, $n = 1, \ldots, l(\delta)$ and $m \in \mathbb{Z}$, $m \equiv \delta \bmod 2$.

On $L^2(i\mathbb{R}^+)$, we denote by $\langle \, , \, \rangle$ the scalar product with respect to the Lebesgue measure. Let $f \in L^2(i\mathbb{R}^+)$, $j \in \{1, \ldots l(\delta)\}$, $m \in \mathbb{Z}$, and $u \in Q_k$. Then

$$\left| \int_0^\infty f(i\tau)(E_j(m, i\tau), E_u) \, d\tau \right| \leq \|f\| \left(\int_{i\mathbb{R}^+} |(E_j(m, i\tau), E_u)|^2 \, d\tau \right)^{1/2};$$

therefore, by (3),

(4)
$$\left| \int_0^\infty f(i\tau)(E_j(m, i\tau), E_u)\, d\tau \right| \leqslant (2\pi)^{1/2} \|f\| \|E_u\|,$$

which means that

(5)
$$\mu_f: E_u \mapsto \int_0^\infty f(i\tau)(E_u, E_j(m, i\tau))\, d\tau$$

is a continuous linear form on E_Q. The latter is dense (17.2), so μ_f extends to a continuous linear form on H_{ct} with norm $\leqslant (2\pi)^{1/2} \|f\|$. As in 13.8, by the Riesz theorem this form represents an element of H_{ct}, to be denoted $\int_0^\infty f(i\tau).E_j(m, i\tau)\, d\tau$.

Consider now the direct integral $\mathbb{H}_{\delta, j}$ (15.9). A K-finite section f is a map

(6)
$$f: i\tau \mapsto \sum_{m \equiv \delta(2)} f_m(i\tau)\varphi_{j, m, i\tau},$$

where the sum over m is finite. The square norm $\|f\|^2$ of that section is $\sum \|f_m\|^2$, and the f_m are square integrable. The jth *Eisenstein transform*

(7)
$$_\delta\mathcal{E}_j: \mathbb{H}_{\delta, j} \mapsto {_\delta}H_{ct} \quad (j = 1, \ldots, l(\delta))$$

is defined by

(8)
$$_\delta\mathcal{E}_j(f) = (2\pi)^{-1/2} \sum_{m \equiv \delta(2)} \int_0^\infty f_m(i\tau).E_j(m, i\tau)\, d\tau.$$

By (4),

(9)
$$|(_\delta\mathcal{E}_j(f), E_u)| \leqslant (2\pi)^{1/2} \|f\| \|E_u\|.$$

17.5   If $f \in C_c^\infty(i\mathbb{R}^+)$, or (slightly more generally) if f is compactly supported and continuously differentiable, then it can be proved directly that $\mathcal{E}_j(f)$ is square integrable using integration by parts, as indicated in 7.1 of [36], with the estimate (6.20) there being replaced by 11.16 for B on the imaginary axis. Note that it follows from (9) that $f \mapsto {_\delta}\mathcal{E}_j f$ is a continuous map of $L^2(i\mathbb{R}^+)$ into H_{ct}.

A section f as in 17.4(5) is said to be compactly supported (resp. smooth, resp. Schwartz) if f is so. (See [46, I, p. 133] for the definition of a Schwartz function. All we need to know is that the restriction of a PW function is a Schwartz function.)

The following proposition (which implies 13.8(8)) differs only in minor details from [36, Prop. 7.1]. The proof is basically the same, and is given here for the sake of completeness.

17.6 Proposition.   *Let* f *(resp. g) be a section of* $_\delta\mathbb{H}_j$ *(resp.* $_\delta\mathbb{H}_k$*). Then*

(1) $$(_\delta\mathcal{E}_j(f), _\delta\mathcal{E}_k(g)) = \delta_{j,k}\langle f, g\rangle \quad (1 \leqslant j, k \leqslant l(\delta)).$$

Since $C_c^\infty(i\mathbb{R}^+)$ is dense in $L^2(i\mathbb{R}^+)$, we may (and do) assume that f and g are compactly supported and smooth.

The sections defined by the f_m (resp. g_m) are orthogonal by construction. Hence, if $j = k$ then the scalar product is

(2) $$\langle f, g\rangle = \sum_m \langle f_m, g_m\rangle.$$

If $j \neq k$ then the scalar product will in this proof be given by the same formula, $_\delta\mathbb{H}_j$ and $_\delta\mathbb{H}_k$ being identified to $_\delta\mathbb{H}$ canonically (see 15.9).

Proof. Two elements of H_{ct} of different K-types are necessarily orthogonal, so we may assume that $f = f_m$ and $g = g_m$ for some given $m \equiv \delta \bmod 2$. Then

(3)
$$_\delta\mathcal{E}_j(f) = (2\pi)^{-1/2} \int_0^\infty f(ir).E_j(m, ir)\,dr,$$
$$_\delta\mathcal{E}_k(g) = (2\pi)^{-1/2} \int_0^\infty g(it).E_k(m, it)\,dt.$$

Fix $T > 0$. For the truncations at T (see 12.16) we also have, as follows from the definition,

(4)
$$_\delta\mathcal{E}_j(f)^T = (2\pi)^{-1/2} \int_0^\infty f(ir).E_j(m, ir)^T\,dr,$$
$$_\delta\mathcal{E}_k(g)^T = (2\pi)^{-1/2} \int_0^\infty g(it).E_k(m, it)^T\,dt;$$

hence

(5) $\left(_\delta\mathcal{E}_j(f)^T, _\delta\mathcal{E}_k(g)^T\right)$

$$= (2\pi)^{-1} \int_0^\infty dr\, f(ir) \int_0^\infty dt\, \overline{g(it)}\left(E_j(m, ir)^T, E_k(m, it)^T\right).$$

In the rest of the proof, we write $E_j(ir)$ and $E_k(it)$ for $E_j(m, ir)$ and $E_k(m, it)$. By the Maass–Selberg relations (12.10),

(6) $$\left(E_j(ir)^T, E_k(it)^T\right) = \frac{T^{-i(r+t)}}{-i(r+t)} c_{j,k}(ir) + \frac{T^i(r+t)}{i(r+t)} \overline{c_{k,j}(it)}$$

$$+ \frac{T^{i(t-r)}}{i(t-r)}\left(\delta_{j,k} - \sum_n c_{j,n}(ir)\overline{c_{k,n}(it)}\right)$$

$$+ \delta_{j,k}\frac{T^{i(t-r)} - T^{-i(t-r)}}{i(t-r)}.$$

The sum $A(r, t, T)$ of the first two terms can be written as

$$(7) \quad A(r, t, T) = T^{i(r+t)} \frac{c_{j,k}(-it) - c_{j,k}(ir)}{i(r+t)} + \frac{T^{i(r+t)} - T^{-i(r+t)}}{i(r+t)} c_{j,k}(ir).$$

Note that $T^{i(r+t)}$ is the derivative with respect to t of $(i \ln T)^{-1}.T^{i(r+t)}$; its coefficient is continuously differentiable on $\mathbb{R}^+ \times \mathbb{R}^+$. The second summand of $A(r, t, T)$ can be written as

$$(8) \quad 2 \sin(\ln T (r + t)).(r + t)^{-1}.c_{j,k}(ir).$$

Integration by parts then shows that

$$(9) \quad \left| \int_0^\infty \overline{g(it)}.A(r, t, T)\, dt \right| \prec (\ln T)^{-1} \quad (T > 0).$$

The matrix C is unitary on the imaginary axis (11.13), so

$$B(r, t, T) = -iT^{i(t-r)}.(t - r)^{-1}.\left(\delta_{j,k} - \sum_n c_{j,n}(ir)\overline{c_{k,n}(it)} \right)$$

is continuously differentiable on $\mathbb{R}^+ \times \mathbb{R}^+$. The derivative of

$$B(r, t, T).T^{i(r-t)}.\overline{g(it)}$$

is bounded on $\mathbb{R}^+$ and tends to zero as $t \to \infty$. Integration by parts again shows that

$$(10) \quad \left| \int_0^\infty B(r, t, T)\overline{g(it)}\, dt \right| \prec (\ln T)^{-1} \quad (T > 0).$$

This implies

$$(11) \quad \left| \int_0^\infty f(ir)\, dr \int_0^\infty \left(A(r, t, T) + B(r, t, T) \right)\overline{g(it)}\, dt \right| \prec (\ln T)^{-1}.$$

There remains to consider the last term in (6) for $j = k$; this term is equal to $2 \sin(\ln T.(t - r))(t - r)^{-1}$. Let $u = t - r$, and consider

$$(12) \quad C(r, T) = \int_{-\infty}^\infty 2 \sin(\ln T.u).\overline{g(i(u + r))}u^{-1}\, du.$$

We have

$$(13) \quad C(r, t) = \overline{g(ir)} \int_{-\infty}^\infty \sin(\ln T.u).u^{-1}\, du$$

$$+ 2 \int_{-\infty}^\infty \sin(\ln T.u).u^{-1}\left(\overline{g(i(u + r))} - \overline{g(ir)} \right) du.$$

The first integral is equal to $2\pi \overline{g(ir)}$. Again using integration by parts, we see that the second is $\prec (\ln T)^{-1}$ in absolute value. Altogether we have

(14) $\quad \left| \left((_\delta\mathcal{E}_j f)^T, (_\delta\mathcal{E}_k g)^T \right) - \int_0^\infty f(ir)\overline{g(i,r)}\,dr \right| \prec (\ln T)^{-1},$

so

(15) $\quad \lim_{T\to\infty} \left((_\delta\mathcal{E}_j f)^T, (_\delta\mathcal{E}_k g)^T \right) = \delta_{j,i} \langle f, g \rangle.$

To end the proof, it suffices to show that

(16) $\quad \lim_{T\to\infty} \left((_\delta\mathcal{E}_j f)^T, (_\delta\mathcal{E}_k g)^T \right) = (_\delta\mathcal{E}_j f, _\delta\mathcal{E}_k g) \quad (1 \leqslant j, k \leqslant l(\delta)).$

In view of the continuity of the scalar product on a Hilbert space, this in turn will follow from

(17) $\quad \| _\delta\mathcal{E}_j f - (_\delta\mathcal{E}_j f)^T \| \prec (\ln T)^{-1/2} \quad (j = 1, \dots, l(\delta)).$

Let $x \in \mathfrak{S}_{i,T}$. The difference $\left(_\delta\mathcal{E}_j(i\tau) - _\delta\mathcal{E}_j(i\tau)^T \right)(x)$ is the ith constant term by definition, so

$$\left(_\delta\mathcal{E}_j f - (_\delta\mathcal{E}_j f)^T \right)(x) = \int_0^\infty f(i\tau) h_{P_i}(x) \{ \delta_{j,i} h_{P_i}^{i\tau}(x) + c_{j,i}(i\tau) h_{P_i}^{-i\tau}(x) \} \, d\tau.$$

As a function of τ, $h_{P_i}^{i\tau}(x)$ is the derivative of $(i \ln h_{P_i}(x))^{-1} h_{P_i}^{i\tau}(x)$; therefore, integration by parts yields

(18) $\quad \left| \left(_\delta\mathcal{E}_j f - (_\delta\mathcal{E}_j f)^T \right)(x) \right| \leqslant C . h_{P_i}(x)(\ln h_{P_i}(x))^{-1}.$

The difference on the left-hand side is zero on the complement of the union of the $\mathfrak{S}_{i,T}$ $(i = 1, \dots, l(\delta))$. It suffices to majorize the square norm of that difference on $\mathfrak{S}_{i,T}$. As in 12.10, set $t = h_{P_i}(x)$ and recall that the restriction to A_i of the Haar measure is $dt . t^{-3}$. Consequently,

$$\| _\delta\mathcal{E}_j f - (_\delta\mathcal{E}_j f)^T \|^2 \leqslant C . \int_T^\infty t^{-1}(\ln t)^{-2} \, dt = C . (\ln T)^{-1}$$

for some constant C, and (17) follows.

17.7 Theorem. *Let $_\delta\mathbb{H}_{l(\delta)}$ be the orthogonal direct sum of the $_\delta\mathbb{H}_j$ $(j \leqslant l(\delta))$. The direct sum $_\delta\mathcal{E}$ of the $_\delta\mathcal{E}_j$ $(j \leqslant l(\delta))$ induces an isomorphism of unitary G-modules of $_\delta\mathbb{H}_{l(\delta)}$ onto $_\delta H_{ct}$.*

Proof. By 17.6, $_\delta\mathcal{E}$ restricted to the sums of Schwartz sections of the summands $_\delta\mathbb{H}_j$ is isometric and, in particular, bounded. We claim that the image of $_\delta\mathcal{E}$ is dense and, more precisely, contains the E_u ($u \in {}_\delta Q$, u right K-finite). To see this, associate to E_u the section of $_\delta\mathbb{H}_j$ given by

(1) $\quad i\tau \mapsto (2\pi)^{-1/2} \sum_{m \equiv \delta(2)} (E_u, E_j(m, i\tau)) \varphi_{j,m,i\tau},$

to be denoted $_\delta \mathcal{G}_j(E_u)$, and let $_\delta \mathcal{G}(E_u) \subset \mathbb{H}_{I(\delta)}$ be the sum of the $_\delta \mathcal{G}_j(E_u)$. This is a Schwartz section, as follows from 12.6. By 13.8(5) and (9),

$$(2) \qquad\qquad (_\delta \mathcal{E} \circ _\delta \mathcal{G})(E_u) = E_u.$$

The sums of Schwartz sections of the $_\delta \mathbb{H}_j$ are dense in $\mathbb{H}_{I(\delta)}$. Restricted to those, $_\delta \mathcal{E}$ is an isometric map with dense image. By continuity, it extends to an isomorphism of unitary G-modules, proving the theorem.

The map $_\delta \mathcal{G}$ is the *Godement transform*. It is defined at first on $E_{\delta Q}$ and is isometric by 17.4(3); $_\delta \mathcal{G}$ thus extends to H_{ct} and, by (2), is the inverse to $_\delta \mathcal{E}$.

17.8 Corollary. *The Godement transform is an isomorphism of unitary G-modules of $_\delta H_{ct}$ onto $_\delta \mathbb{H}_{I(\delta)}$. In particular, $_\delta \mathcal{G}_j(E_Q)$ is dense in $_\delta \mathbb{H}_j$ ($j \in I(\delta)$).*

If it were possible to establish the last assertion directly, then the proof of the isomorphism of $_\delta H_{ct}$ with a direct sum of direct integrals could also be established by means of the Godement transform.

What we have called the Godement transform is obtained by analytic continuation and integration from an intertwining operator of V_K to $I(P_j, \delta, s)$, defined at first for $\mathcal{R}s > 1$. To complete the discussion, we give some more details about Godement's construction [23; 24].

17.9 The Godement transform.   For $f \in C_c^\infty(\Gamma \backslash G)$, we formally define a Laplace transform $L_P(f) = \hat{f}$ on $G \times \mathbb{C}$ by

$$(1) \qquad\qquad \hat{f}(g, s) = (2\pi)^{-1/2} \int_A f_P(a.g) h_P(a)^{s-1} \, da.$$

Note that we have the constant term f_P in the integrand. This function does not have compact support, so it is not obvious that the integral converges as was clear in the case of the Fourier–Laplace transform in 12.4. This definition presupposes the choice of (P, A). If this needs to be stressed, we write $_P\hat{f}$ for $\hat{f}$.

Lemma.   *Let C be a symmetric compact subset of G. Then the integral on the right-hand side of (1) converges absolutely and uniformly for $g \in C$ if $\mathcal{R}s > 1$.*

Proof. Let D be a compact subset of G such that $\operatorname{supp} f \subset \Gamma.D$. Then

$$\operatorname{supp} f_P \subset N.\Gamma.D$$

and $f(a.g)$ ($g \in C$) can be nonzero only if $a \in A \cap N.\Gamma D.C$. Let E be a fundamental domain for Γ_N in N. The previous requirement on a is equivalent to $E.a \cap \Gamma.D.C \neq \emptyset$.

As in Section 3, let A_t $(t > 0)$ be the set of $a \in A$ for which $h_P(a) \geqslant t$. Then $E.A_t$ is contained in a Siegel set (3.15). By 3.16, there are only finitely many $\gamma \in \Gamma$ such that $\gamma(D.C) \cap E.A_t$ is not empty. Each such intersection is relatively compact in A, so there exists $c > 0$ such that $f_P(a.g) = 0$ for $g \in C$ and $a \in A_c$. Let us again use $h_P(a) = t$ as the coordinate on A. Then $da = dt/t$ and the integral on the right-hand side of (1) is majorized, independently of $g \in C$, by

$$\int_0^c \frac{dt}{t^{2-\sigma}} \quad (\sigma = \mathcal{R}s),$$

which converges if $\sigma > 1$.

It is clear from the definition that

(2) $(r_g.f)^\wedge = r_g(\hat{f}) \quad (g \in G),$

(3) $(\hat{f})_m = (f_m)^\wedge \quad (m \in \mathbb{Z}).$

Remark. In this definition, f enters only through its constant term. We could equally well start from a function $f \in {}_\delta C_c(N\backslash G)$. Then $f_P = f$ and the support of f is contained in $N.D$, where D is a compact subset of G; this can be written $\Gamma_N.E.D$, where E is a fundamental domain for Γ_N in N. The preceding proof also shows that $L_P(f)$ is well-defined for $\mathcal{R}s > 1$.

Fix $\sigma > 1$. For a fast decreasing function v on $\sigma + i\mathbb{R}$, define

(4) $\check{v}(a) = (2\pi)^{-1/2} \int_{\sigma+i\mathbb{R}} v(s)h_P^{1-s}(a)\, d\tau.$

Then we have an inversion formula

(5) $f(a.g) = (\hat{f}(g, s))^\vee(a).$

To see this, let $\mathcal{F}$ and $\tilde{\mathcal{F}}$ be (respectively) the usual Fourier transform, and inverse Fourier transform so that $\mathcal{F}.\tilde{\mathcal{F}} = \mathrm{Id}$. Then

$$\hat{f}(g, s) = \mathcal{F}(r_g f.h_P^{\sigma-1}) \quad \text{and} \quad \check{v}(a) = h_P(a)^{1-\sigma}\mathcal{F}(v)(a);$$

hence

$$(\hat{f}(g, s))^\vee(a) = h_P(a)^{1-\sigma}\mathcal{F}\tilde{\mathcal{F}}(r_g f.h_P^{\sigma-1})(a) = r_g f(a) = f(a.g).$$

17.10   We now claim that the map $f \mapsto L_P(f)$ defines for $\mathcal{R}s > 1$ an intertwining operator

(1) $\mathbb{L}_{P,s} \colon {}_\delta C_c^\infty(\Gamma\backslash G) \to I(P, \delta, -s).$

Let $p = n.b$ $(n \in N, b \in A)$. Then

$$\hat{f}(p.g, s) = (2\pi)^{1/2} \int_A f_P(a.n.b.g).h_P(a)^{s-1}\, da.$$

We have

$$f_P(a.n.b.g) = f_P(^a n.a.b.g) = f_P(a.b.g).$$

Set $a' = a.b$. Then

$$\int_A f_P(a.n.b.g).h_P(a)^{s-1}\, da = h_P(b)^{1-s}\int_A f_P(a'.g).h_P(a')^{s-1}\, da',$$

(2) $\hat{f}(p.g.s) = h_P(p)^{1-s}.\hat{f}(g,s) \quad (p \in P^0).$

Let ε be the nontrivial central element of G. Then $f(\varepsilon.x) = (-1)^\delta f(x)\ (x \in G)$ by assumption. The same is then true for f_P and $\hat{f}$, so $\hat{f}$ belongs to $I(P,\delta,-s)$. As pointed out in 17.7(2), $L_P: f \to \hat{f}$ commutes with right translations on G. This proves our claim.

17.11 Proposition. *Let $u \in {}_\delta C_c^\infty(\Gamma_P.N\backslash G)$. Let E_u be the associated E-series (12.0, 12.1), and let $m \in \mathbb{Z}$.*

   (i) *The function $s \mapsto L_P(E_{u,m})(1,s)$ has a meromorphic continuation to $\mathbb{C}$, which is holomorphic at every point where $E(P',m,\bar{s})$ is antiholomorphic, in particular in $\Omega = \{s \in \mathbb{C},\ \mathcal{R}s \geq 0,\ s \notin (0,1]\}$. At those points,*

(1) $(2\pi)^{1/2}.L_{P'}(E_{u,m})(1,\bar{s}) = (E_u, E(P',m,s)).$

If u is right K-finite, then $L_{P'}(E_u)(g,s)$ has a meromorphic continuation to $\mathbb{C}$ that is holomorphic on Ω, belongs to $I(P',\delta,-s)_K$ for $s \in \Omega$, and is given by

(2) $(2\pi)^{1/2}.L_{P'}(E_u)(g,s) = \sum_m L_{P'}(E_{u,m})(1,s)\varphi_{P',m,-s} \quad (g \in G,\ s \in \Omega).$

   (ii) *Assume that $E(P',m,s)$ has a simple pole at $z \in (0,1]$, and let $E'(P',m,z)$ be its residue. Then $L_{P'}(E_{u,m})(1,s)$ has at most a simple pole at z, with residue $L_{P'}(E_{u,m})'(1,z)$ given by*

(3) $(2\pi)^{1/2}L_{P'}(E_{u,m})'(1,z) = (E_u, E'(P',m,z)).$

If u is right K-finite then $L_{P'}(E_u)(g,s)$ has at most a simple pole at z, and its residue $L_{P'}(E_u)'(g,z)$ belongs to $I(P',\delta,-z)_K$; moreover,

(4) $L_{P'}(E_u)'(g,z) = \sum_m L_{P'}(E_{u,m})'(1,z)\varphi_{P',m,-z}(g) \quad (g \in G).$

Proof. We first establish (1) for $\mathcal{R}s > 1$. The function E_u is a convergent sum of its components $E_{u,n}\ (n \in \mathbb{Z})$, of which all are orthogonal to $E(P',m,s)$ except possibly for $n = m$, so

(5) $(E_u, E(P',m,s)) = (E_{u,m}, E(P',m,s)).$

By 12.3,

(6) $$(E_u, E(P', m, s)) = (E_{u,P'}, \varphi_{P',m,s})_{N'\backslash G}.$$

The quotient $N'\backslash G$ is diffeomorphic to $A' \times K$, and the invariant quotient Haar measure is $h_{P'}(a')^{-2}.da'.dk$. Therefore,

$$(E_u, E(P', m, s)) = \int_{A' \times K} h_{P'}(a')^{-2}.E_{u,P'}(a'.k)h_{P'}(a')^{1+\bar{s}}\chi_{-m}(k)\, da'\, dk$$

$$= \int_{A'} E_{u,P',m}(a')h_{P'}(a')^{\bar{s}-1}\, da'.$$

We have $E_{u,P',m} = E_{u,m,P'}$, since taking the constant term commutes with right translations. Hence (1) is proved for $\mathcal{R}s > 1$.

By 11.9 and 11.13, the right-hand side of (1) has a meromorphic continuation to $\mathbb{C}$ and is antiholomorphic at every point where $E(P', m, s)$ is holomorphic. This provides the analytic continuation of the left-hand side. For $\mathcal{R}s > 1$, the function $L_{P'}(E_{u,m})(g, s)$ is of right K-type m in the principal series $I(P', \delta, -s)$, by 17.10, and is therefore a multiple of $\varphi_{P',m,-s}$. The latter is equal to 1 at 1, so

(7) $$L_{P'}(E_{u,m})(g,s) = L_{P'}(e_{u,m})(1, s).\varphi_{P',m,-s}(g) \quad (g \in G).$$

The function $\varphi_{P',m,-s}$ is entire in s, so this provides the meromorphic continuation to $\mathbb{C}$, holomorphic on E. If u is right K-finite then $E_{u,m}$ is nonzero for only finitely many m, and (2) follows.

Equations (3) and (4) are obtained by multiplying both sides of (1) and (2) by $(s - z)$ and letting $s \to z$.

Remark. The proposition implies in particular that $L_{P'}(E_{u,m})(1, s)$ is holomorphic at $z \in (0, 1]$ if either $E(P', m, s)$ is holomorphic at z or $E(P', m, s)$ has a simple pole and $(E_u, E'(P', m, z)) = 0$.

17.12 Comparisons. In 12.4, we introduced a Fourier–Laplace transform by setting, for $u \in C_c^\infty(A)$,

(1) $$\hat{u}(s) = \int_A u(a).h_P(a)^{-(1+s)}\, da,$$

which is defined for all $s \in \mathbb{C}$. In order to avoid confusion, we use here only the notation $L_P(f)$ for the Godement transform of 17.9. By extending the definition of $L_P(f)$ as in the remark in 17.9, we see that

(2) $$\hat{u}(s) = L_P(u)(1, -s) \quad (u \in {}_\delta C_c^\infty(N\backslash G), \mathcal{R}s > 1).$$

From 12.6 and 17.11 we obtain

(3) $$L_{P'}(E_u)(1, -s) = \{\delta_{P',P}\hat{u}(-\bar{s}) + \overline{c_{P'|P}(s)}\hat{u}(\bar{s})\},$$
$$L_{P'}(E_u)(1, \bar{s}) = \{\delta_{P',P}\hat{u}(-\bar{s}) + \overline{c_{P'|P}(s)}\hat{u}(\bar{s})\}.$$

Alternatively, by 12.6(2) we have

(4) $$L_{P'}(E_u)(1, s) = \{\delta_{P', P} \hat{u}(-s) + c_{P'|P}(-s)\hat{u}(s)\},$$

which shows that the analytic continuation of $c_{P|P'}(s)$ implies that of $L_{P'}(E_u)(1, s)$ and, in fact, that any pole of the left-hand side is one of $c_{P|P'}(s)$.

Similarly, 12.6(3) and 17.11(3) imply

(5) $$L_{P'}(E_u)'_m(1, z) = \hat{u}(z)c'_{P'|P}(z).$$

17.13 Recall that $\varphi_{P,m,s} \mapsto \varphi_{P,m,-s}$ yields an isomorphism μ_s of $I(P, \delta, -s)$ onto $I(P, \delta, s)$. Let $\tilde{L}_P = \mu_s \circ L_P$. Then the Godement transform $_\delta\mathcal{G}_j$ defined in the course of the proof of 17.7 associates to E_u the section

(1) $$i\tau \mapsto \sum_m \tilde{L}_{P_j}(E_u)(\cdot, i\tau).$$

In view of 17.11(1), this coincides with the definition in 17.7(1).

18 Concluding remarks

The spectral decomposition ends this description of some basic material on the analytic theory of automorphic forms on $SL_2(\mathbb{R})$. This is, however, only the starting point of the theory. We now add some comments on other developments, mainly for orientation and to indicate some literature for further study, without aiming at completeness.

18.1 Let $\mathcal{A}(s, m)$ be the space of automorphic forms of right K-type m that are eigenfunctions of $\mathcal{C}$ with eigenvalue $(s^2 - 1)/2$. By definition, this space is the orthogonal direct sum of the subspace $°\mathcal{A}(s, m)$ of cusp forms and of its orthogonal complement $\mathcal{A}_1(s, m)$. In Section 12, we obtained some information on the latter: its dimension is equal to the number of cusps or neat cusps; it is generated by Eisenstein series $E(r, m)$ holomorphic at s and suitable limits of Eisenstein series. Although this description is not quite exhaustive, depending notably on the poles of Eisenstein series, it is quite substantial, so that the main remaining issue is the determination of the cusp forms. Note that, in the cocompact case, $\mathcal{A}_1(s, m) = \{0\}$ and so nothing has been achieved toward the description of $\mathcal{A}(s, m)$.

By 16.2, the cuspidal spectrum $°L^2(\Gamma \backslash G)$ is a Hilbert direct sum of irreducible unitary G-modules $\pi \in \hat{G}$ with finite multiplicities, say $m(\pi, \Gamma)$. We know the K-types of all irreducible unitary representations of G (see §15), and they all have multiplicity 1 (a very special property of $SL_2(\mathbb{R})$). Therefore dim $°\mathcal{A}(s, m)$ is the sum of the $m(\pi, \Gamma)$, where π runs through the $\pi \in \hat{G}$ for which the eigenvalue $\mathcal{C}(\pi)$ of $\mathcal{C}$ is $(s^2 - 1)/2$ and which contain the K-type m. From the discussion of K-types in irreducible representations in Section 15, we see that this depends only on the parity of m. For a given m, we are in fact dealing with the spectral decomposition of the space of cusp forms of K-type m with respect to $\mathcal{C}$. If $m = 0$, this is the same as the spectral decomposition with respect to the Laplace–Beltrami operator Δ.

18.2 If π is a discrete representation D_n ($n \in \mathbb{Z}$, $n \neq 0$) (see 15.6 and 15.10), and especially if it is integrable (i.e., if $|n| \geqslant 2$), then the construction of holomorphic (or antiholomorphic) Poincaré series (see 6.2) yields cusp forms (8.9). If such a series is not identically zero, this shows (16.3) that $m(D_n, \Gamma) \neq 0$. Classically, the holomorphic automorphic forms of weight m can be viewed as holomorphic sections of some line bundles; much information on such sections can be derived from the Riemann–Roch theorem, applied to $\Gamma \backslash X$ if it is compact or to its natural compactification $\Gamma \backslash X^*$ (3.19) otherwise (see [53, II] or [51]). Geometrically or analytically, it can be seen that D_n will occur in the spectrum of a subgroup of Γ of sufficiently large (finite) index.

18.3 The main outstanding problem then is to determine $m(\pi, \Gamma)$ when π belongs to the unitary principal series or to the complementary series. Recall that $C = -2.\Delta$, where Δ is the Laplace–Beltrami operator. Thus the eigenvalue $\Delta(\pi)$ of Δ can be written $(1 - s^2)/4$. If π is in the complementary series then $\Delta(\pi) \in (0, 1/4)$. If π belongs to the principal unitary series then $s = ir$, with r real and

$$\Delta(\pi) = (1 + r^2)/4 \geqslant 1/4.$$

Consider first the case of the unitary principal series. For $R \in \mathbb{R}$, let $N(\Gamma, R)$ be the number of cusp forms corresponding to eigenvalues in $[1/4, R]$. Let $\Gamma = \Gamma(n)$ be a congruence subgroup (3.20.1). Using his trace formula, Selberg showed that $N(\Gamma, R) \to \infty$ and also gave its asymptotic behavior [36, 11]. This led to the conjecture that $N(\Gamma, R)$ should be unbounded for any Γ. However, Phillips and Sarnak [45] proved that the cuspidal forms for a given Γ may disappear under a small deformation of Γ. This, as well as other results, gave some reason to believe that, on the contrary, $N(\Gamma, R)$ might remain bounded for a "generic" Γ (see [47] for a discussion and references, and [60] for a more recent contribution). As to the complementary series, a famous conjecture of Selberg asserts that, for congruence subgroups, all eigenvalues of Δ are at least $1/4$; Selberg himself proved this for the modular group. In principle, the complementary series can occur as residues of Eisenstein series (16.6), but this is known not to happen for congruence subgroups (see [36, 11.3]). So the remaining question is whether the complementary series appears in the cuspidal spectrum of a congruence subgroup. For other Γ, it may (see [49] for a survey).

18.4 Selberg obtained his results on cusp forms as an application of his trace formula. The latter is the main tool in analytic theory and would be the next topic in a comprehensive exposition. However, to include an adequate treatment might easily increase the size of this book by a half, at least – a prospect the author found too daunting, especially in view of the abundant literature on the trace formula and its manifold applications. See notably [50] and other papers in the same volume, the books by D. A. Hejhal [32, 33], the introductory article [20] and the books [36], [18], and [19].

18.5 The central subject matter in this book is a notion of automorphic form generalizing the classical notion of holomorphic automorphic form. But already in that case, other extensions had been introduced. Let (r, E) be a finite dimensional unitary representation of Γ. Then, instead of Γ-invariant functions, consider smooth maps $f: G \to E$ satisfying the condition

$$f(\gamma.g) = r(\gamma).f(g) \quad (\gamma \in \Gamma, \, g \in G),$$

that is, elements of the induced representation $I_\Gamma^G(E)$. This is, for instance, the framework of [50]. Next one can consider forms of rational, or even arbitrary real, weight. Geometrically, this amounts to replacing G by a finite covering (for rational weights) and or by its universal covering (for arbitrary real weights). The case of half-integral weight, associated to a twofold covering, is particularly important. The fundamental group of $SL_2(\mathbb{R})$ is $\mathbb{Z}$, so G has an n-fold covering G_n for any n. The automorphic forms of weight a/n ($a \in \mathbb{Z}$) are in principle attached to discrete subgroups of G_n. However, the proper principal congruence subgroups are known to be free (in fact, all normal nontrivial subgroups of the modular group are free, except for two, according to [44]) and hence can be lifted to any covering of G. So for them or any subgroup of finite index, automorphic forms of weight a/n could be defined directly, without reference to G_n, in terms of multipliers, as was done classically.

For the applications to algebraic geometry or arithmetic, it is necessary to go over to $GL_2(\mathbb{R})$ and also to "adelize". In fact, some of the references given previously operate in that framework. For this, the reader may also consult the Proceedings of the Summer School in Antwerp [38; 15; 39; 3] as well as [18] and [20]. One must also mention the most spectacular recent advance, the proof by A. Wiles (with some assist by R. Taylor) of the Shimura–Tanyiama conjecture for semistable elliptic curves over $\mathbb{Q}$, with a proof of Fermat's "last theorem" as a consequence (see [55; 59]).

18.6 The foregoing all pertains solely to SL_2 or GL_2. There is also an immense literature on the theory for more general semisimple or reductive groups. As already pointed out, our presentation has been much influenced by the treatment of the general case in [31] or [41]. See also [43] for an adelic version and [1] for an introduction to it. Much of it is geared toward the Langlands program, to which [9] (including Part 2) is still the most complete general introduction.

References

[1] J. Arthur, "Eisenstein series and the trace formula", in [9], pp. 253–74.

[2] W. Baily, *Introductory lectures on automorphic forms,* Publ. Math. Soc. Japan **12** (1973), I. Shoten and Princeton Univ. Press.

[3] B. J. Birch and W. Kuyk (eds.), *Modular functions of one variable, IV* (Proc. Internat. Summer School, Univ. Antwerp, 17 July – 3 August, 1972), Lecture Notes in Math., **476**, Springer-Verlag, Berlin, 1975.

[4] A. Borel, "Introduction to automorphic forms", Proc. Sympos. Pure Math., **9**, pp. 199–210, Amer. Math. Soc., Providence, RI, 1966.

[5] A. Borel, "Ensembles fondamentaux pour les groupes arithmétiques et formes automorphes", Cours à l'I.H.P. 1964 (notes par H. Jacquet, J.-J. Sansuc, et B. Schiffman), Secrétariat Math. E.N.S., 1967.

[6] A. Borel, *Introduction aux groupes arithmétiques,* Hermann, Paris, 1969.

[7] A. Borel, *Représentations des groupes localement compacts,* Lecture Notes in Math., **276**, Springer-Verlag, Berlin, 1972.

[8] A. Borel, "Introduction to automorphic forms in one variable", Adv. in Math (China) **25** (1996), 97–158.

[9] A. Borel and W. Casselman (eds.), *Automorphic forms, representations and L-functions,* Proc. Sympos. Pure Math., **33**, part I, Amer. Math. Soc., Providence, RI, 1977.

[10] N. Bourbaki, *Intégration,* chap. 7, 8, Hermann, Paris, 1963.

[11] N. Bourbaki, *Topologie générale I,* nouvelle édition, Hermann, Paris, 1971.

[12] N. Bourbaki, *Groupes et algèbres de Lie I,* Hermann, Paris, 1971.

[13] N. Bourbaki, *Variétés différentielles et analytiques. Fascicule de résultats,* §§1 à 7, Hermann, Paris, 1972.

[14] N. Bourbaki, *Fonctions d'une variable réelle,* Hermann, Paris, 1976.

[15] P. Deligne and W. Kuyk (eds.), *Modular functions of one variable, II* (Proc. Internat. Summer School, Univ. Antwerp, 17 July – 3 August, 1972), Lecture Notes in Math., **349**, Springer-Verlag, Berlin, 1973.

[16] J. Dixmier and P. Malliavin, "Factorisations de fonctions et de vecteurs indéfiniment différentiables", Bull. Sci. Math. (2) **102** (1978), 307–30.

[17] M. Duflo and J.-P. Labesse, "Sur la formule des traces de Selberg", Ann. Sci. École Norm. Sup. (4) **4** (1971), 193–284.

[18] S. Gelbart, *Automorphic forms on Adele groups,* Ann. of Math. Stud., **83**, Princeton Univ. Press, Princeton, NJ, 1975.

[19] S. Gelbart, *The Arthur–Selberg trace formula,* Univ. Lecture Ser., **9**, Amer. Math. Soc., Providence, RI, 1996.

[20] S. Gelbart and H. Jacquet, "Forms of GL(2) from the analytic point of view", in [9], pp. 213–51.

[21] I. M. Gelfand, M. I. Graev, and I. I. Piatetski-Shapiro, *Representation theory and automorphic functions*, Saunders, Philadelphia, 1969.

[22] R. Godement, "Série de Poincaré et Spitzenformen", Sém. H. Cartan 1957/58, exp. **10**.

[23] R. Godement, "Analyse spectrale des fonctions modulaires", Sém. Bourbaki 1964/65, exp. **278**.

[24] R. Godement, "The decomposition of $L^2(\Gamma\backslash G)$ for $\Gamma = \mathrm{SL}_2(\mathbb{Z})$", Proc. Sympos. Pure Math., **9**, pp. 211–24, Amer. Math. Soc., Providence, RI, 1966.

[25] R. Godement, "Introduction à la théorie de Langlands", Sém. Bourbaki 1966/67, exp. **321**, Benjamin, 1968.

[26] W. Gröbner and N. Hofreiter, *Integraltafel II*, Springer-Verlag, Vienna, 1961.

[27] G. H. Hardy and E. M. Wright, *An introduction to the theory of numbers*, 5th ed., Clarendon, Oxford, 1979.

[28] Harish-Chandra, "Representations of a semisimple Lie group on a Banach space, I", Trans. Amer. Math. Soc. **75** (1953), 185–243.

[29] Harish-Chandra, "Automorphic forms on a semisimple Lie group", Proc. Nat. Acad. Sci. USA **45** (1959), 570–73.

[30] Harish-Chandra, "Discrete series for semi-simple Lie groups, II", Acta Math. **166** (1966), 1–111.

[31] Harish-Chandra, *Automorphic forms on semisimple Lie groups* (notes by G. J. M. Mars), Lecture Notes in Math., **62**, Springer-Verlag, Berlin, 1968.

[32] D. A. Hejhal, *The Selberg trace formula for* PSL(2, *R*), *I*, Lecture Notes in Math., **548**, Springer-Verlag, Berlin, 1976.

[33] D. A. Hejhal, *The Selberg trace formula for* PSL(2, **R**), *II*, Lecture Notes in Math., **1001**, Springer-Verlag, Berlin, 1983.

[34] M. Hervé, "Fonctions automorphes d'une variable. Etude des points paraboliques", Sém. H. Cartan 1953/54, chap. III.

[35] L. Hörmander, *Linear partial differential operators*, Grundlehren Math. Wiss., **116**, Springer-Verlag, New York, 1963.

[36] H. Iwaniec, *Introduction to the spectral theory of automorphic forms*, Rev. Mat. Iberoamericana, Madrid, 1995.

[37] A. Knapp, *Representation theory of semisimple groups*, Princeton Math. Ser., **36**, Princeton Univ. Press, Princeton, NJ, 1986.

[38] W. Kuyk (ed.), *Modular functions of one variable, I* (Proc. Internat. Summer School, Univ. Antwerp, 17 July – 3 August, 1972), Lecture Notes in Math., **320**, Springer-Verlag, Berlin, 1973.

[39] W. Kuyk and J-P. Serre (eds.), *Modular functions of one variable, III* (Proc. Internat. Summer School, Univ. Antwerp, 17 July – 3 August, 1972), Lecture Notes in Math., **350**, Springer-Verlag, Berlin, 1973.

[40] S. Lang, $\mathrm{SL}_2(\mathbf{R})$, Addison-Wesley, Reading, MA, 1975.

[41] R. P. Langlands, *On the functional equations satisfied by Eisenstein series*, Lecture Notes in Math., **544**, Springer-Verlag, Berlin, 1976.

[42] H. Maass, "Über eine neue Art von nichtanalytischen automorphen Funktionen und die Bestimmung Dirichletscher Reihen durch Funktionalgleichungen", Math. Ann. **121** (1949), 141–83.

[43] C. Moeglin and J. L. Waldspurger, *Spectral decomposition and Eisenstein series*, Cambridge Tracts in Math., **113**, Cambridge Univ. Press, 1995.

[44] M. Newman, "Free subgroups and normal subgroups of the modular group", Illinois J. Math. **8** (1964), 262–65.

[45] R. S. Phillips and P. Sarnak, "On cusp forms for co-finite subgroups of PSL(2, **R**)", *Invent. Math.* **80** (1985), 339–64.

[46] M. Reed and B. Simon, *Methods of modern mathematical physics I*, rev. ed., and *Methods of modern mathematical physics II*, Academic Press, New York, 1980 and 1975.

[47] P. Sarnak, "On cusp forms", in *The Selberg trace formula and related topics*, Contemp. Math., **53**, pp. 393–407, Amer. Math. Soc., Providence, RI, 1986.

[48] P. Sarnak, *Some applications of modular forms*, Cambridge Tracts in Math., **99**, Cambridge Univ. Press, 1990.

[49] P. Sarnak, "Selberg's eigenvalue conjecture", Notices Amer. Math. Soc. **42** (1995), 1272–77.

[50] A. Selberg, "Harmonic analysis", in *Collected papers, I*, pp. 626–74, Springer-Verlag, Berlin, 1992.

[51] J-P. Serre, "Fonctions automorphes d'une variable: application du théorème de Riemann–Roch", Sém. H. Cartan 1953/54, chap. IV, V.

[52] J-P. Serre, *A course in arithmetic*, Graduate Texts in Math., **7**, Springer-Verlag, Berlin, 1973.

[53] G. Shimura, *Introduction to the arithmetic theory of automorphic functions*, Publ. Math. Soc. Japan, **11**, I. Shoten and Princeton Univ. Press, 1971.

[54] K. Takeuchi, "Commensurability classes of arithmetic triangle groups", J. Fac. Sci. Univ. Tokyo Sect. IA Math. **24** (1977), 201–12.

[55] R. Taylor and A. Wiles, "Ring-theoretic properties of certain Hecke algebras", Ann. of Math. (2) **141** (1995), 553–72.

[56] D. Vogan, *Representations of real reductive Lie groups*, Progr. Math., **15**, Birkhäuser-Boston, Boston, MA, 1981.

[57] N. Wallach, *Real reductive groups, I* and *Real reductive groups, II*, Pure Appl. Math., **132**, Academic Press, Boston, 1988 and 1992.

[58] F. Warner, *Foundations of differentiable manifolds and Lie groups*, Scott Foresman, Glenview, IL, 1971.

[59] A. Wiles, "Modular elliptic curves and Fermat's last theorem", Ann. of Math. (2) **141** (1995), 443–551.

[60] S. Wolpert, "Disappearance of cusp forms in special families", Ann. of Math. (2) **139** (1994), 239–91.

[61] K. Yosida, *Functional analysis* (6th ed.), Grundlehren Math. Wiss., **123**, Springer-Verlag, Berlin, 1980.

Notation index

The notation introduced in 1.5 to 1.12 is not repeated here.

Subject index